Relationship Obits

EDITED AND COMPILED BY

Kathleen Horan

CREATOR OF RELATIONSHIPOBIT.COM

HarperOne
An Imprint of HarperCollins*Publishers*

Relationship Obits

The final resting place for love gone wrong

Lyrics to "I Don't Want to Get Over You" by Stephin Merritt. Published with permission of the writer.

HarperCollins books may be purchased for educational, business, or sales promotional use. For information please write: Special Markets Department, HarperCollins Publishers, 10 East 53rd Street, New York, NY 10022.

HarperCollins Web site: http://www.harpercollins.com

Book design and typography by Ralph Fowler / rlf design

Illustrations by Maura J. Zimmer

FIRST EDITION

Library of Congress Cataloging-in-Publication Data is available upon request.

ISBN 978–0–06–173516–5

09 10 11 12 13 RRD(H) 10 9 8 7 6 5 4 3 2 1

For the bereaved . . .

. . . and to Esther,

for shining a light.

Contents*

---

*Arranged by cause of death

The Death of True Love

Doomed from the Start

I F***ed Up

I don't want to get over you. I guess I could take
a sleeping pill and sleep at will and not have to
go through what I go through. I guess I should take
Prozac, right, and just smile all night at somebody new,
somebody not too bright but sweet and kind who would
try to get you off my mind. I could leave this agony behind
which is just what I'd do if I wanted to, but I don't
want to get over you cause I don't want to get over love.
I could listen to my therapist, pretend you don't exist
and not have to dream of what I dream of; I could listen
to all my friends and go out again and pretend it's enough,
or I could make a career of being blue—I could dress
in black and read Camus, smoke clove cigarettes and drink
vermouth like I was 17 that would be a scream but I
don't want to get over you.

—The Magnetic Fields
(From the song, "I Don't Want To Get Over You")

Introduction

When you go through a breakup, basically all you have are breakup songs. Not that you can really listen when you're in the acute phase. (The acute phase: skin yanked off, crushed-guts-oozing-out-of-your-sweater kind of heartache.) If you've gone through it, I don't have to explain. One of the worst parts is you feel like a cliché. Your friends want to help but there's not much that can be done—invariably you're steered towards more ice cream or hooking up with someone new. After a few weeks you're supposed to be all healed up. Neither you nor your friends will admit that it's gonna take a big-ass chunk of time for you to get over losing a real love. (Or in some cases, an infatuation.) Especially if someone breaks up with you and you don't see it coming.

That's pretty much what happened with me. Except that two weeks after my boyfriend and I broke up, my dad died. I flew out to California to help make arrangements and grieve. As I wrote my father's obituary, I was surprised by how oddly comforting it was because I sort of saw his life more in focus—if that makes sense—with a beginning, middle, and end there like a real story. I've always appreciated obituaries and have prepared audio obits for my job as a radio reporter. But of course, it's different when it happens to you. "It" meaning loss. That's another thing that was so surprising to me—experiencing the breakup and death so close together—that they felt incredibly similar. Of course one person was my father—no one could ever replace him but it was that feeling of shock—and of having someone you love very deeply basically wiped out.

I was also struck that the "stages of grief" written about by Dr. Elisabeth Kubler-Ross—denial, anger, bargaining, depression, and acceptance—are also clearly felt during a breakup. It turns out the only difference is that with your ex, you can still run into them picking up your drycleaning.

And the thing with "real" death is that you usually get time off from work and life to grieve. People send condolence cards and you attend a memorial or have some type of ceremony to say goodbye. The process helps tons. That kind of separation from business-as-usual helps you reorient and discover what might come next.

When I got home from my dad's service, I thought that packing up every particle that reminded me of the diabolical nerd boyfriend and sealing it into a Hershey's Syrup box would help me be able to finally get on with things, but it really didn't. As I gathered up the letters, cards, and gifts he had given me, I was catapulted into the life of the relationship. And I remembered the obituary writing process that I went through for my dad—how it made his life feel real, even in death, and helped me let him go too.

So, I started writing another obituary, this time for my ex and me. I laid out the facts just like I did before—what were the most memorable days (and nights), the cause of death, and the survivors? Part of me knew that it was a goofy thing to do, but up until then nothing felt as helpful. I could memorialize the thing before I moved on. I spent nearly three years of my life with this guy and documenting it made the stinging pain I was still feeling seem less senseless. I didn't have to look to him for reasons why it ended—as I wrote, they became clear. It was also a way to experience the love again without feeling like a ninny for getting hurt.

After I wrote the obituary, I wondered if other people might also be drawn to writing about their break-ups in this way.

As people do these days, I thought to conduct my social experiment on the Internet and I hooked up with a Web designer who helped me create an obituary page (similar to a newspaper) for "dead relationships." Relationshipobit.com was born.

We launched on February 13th (last chance to write an obit before Valentine's Day) and called the party "A Wake for Love."

About 20 obits were on the site on the 13th. Within 24 hours we had hundreds. And they keep coming . . .

In the months since, I have been awed by the obituaries people have written and posted.

Hearts that still sting from 40-year-old breakups, unions dissolved over the bite of a hamburger, a Coors Light, or narcissism "stage 4."

Many have written from all over the world to say that the site has provided them some kind of closure.

And if they're not currently going through a breakup and have forgotten how bad it can be (like childbirth), some are surprised by the zillions of ways people find themselves together and then apart.

You'll find an array of them assembled here. The tragic, the funny, and the "I thought my ex sucked but he didn't cheat on me while I had cancer" kind of obits. Some names and identifying characteristics of the "obitees" have been altered to protect their identities—but the tales of love's mortality remain true.

If you find yourself currently schlepping around the pain of a breakup, this book is dedicated to you.

—Kathleen Horan,
November 2008

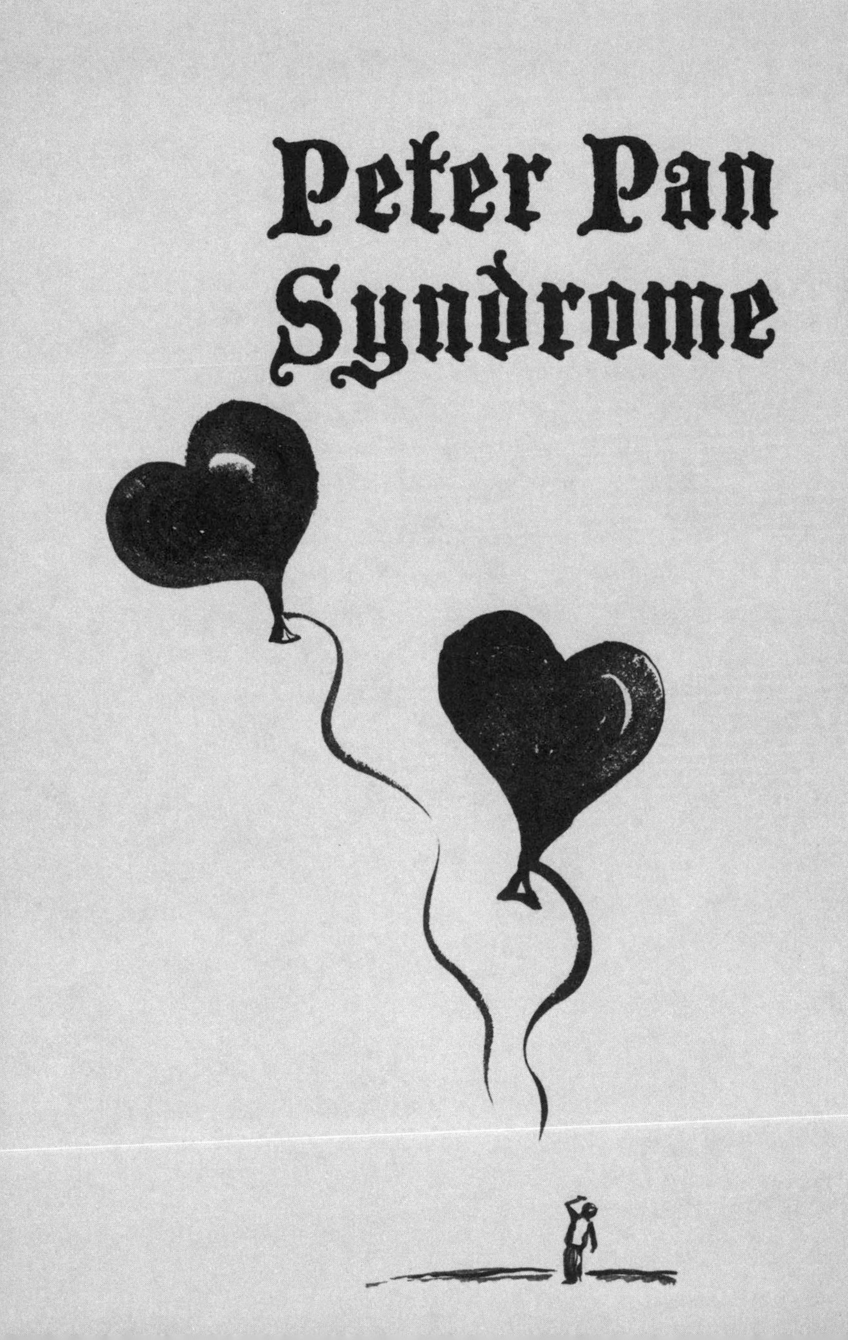
Peter Pan
Syndrome

"This relationship died late Wednesday night."

Cause of Death:
Unknown

Kasey and Travis
Born: August 29, 2008
Died: January 7, 2009

This relationship died late Wednesday night. Doctors are stumped as to the cause, though they've released a statement that it may have suffered from a terminal illness.

Famous Last Words:
I guess that's it then.

Cause of Death:

Inabilility to commit

Merry Xmas

Carole and Gil

Born: December 12, 1992

Died: December 17, 2007

Fifteen years. Fifteen YEARS! And he breaks up with me by leaving a voicemail on my work phone. And a week before Christmas, no less . . .

Met: Through a mutual friend.

What I'll remember: The life we built together. In fifteen years you build a routine. Now it's gone and I feel lost.

What I won't miss: Having to pretend everything's ok, his cowardly ways, his selfishness, his inability to really commit. There was always distance and he put it there.

There are no survivors.

No kids. He robbed me of any opportunity for a family.

Famous Last Words:

Sex is not everything.

"And he breaks up with me by leaving a voicemail on my work phone."

"He said the relationship needed to be snuffed because it was soot on the wings of his career."

Cause of Death: Strangulation by Peter Pan tights

H and D
Born: May 18, 2004
Died: November 28, 2006

The couple with the squinty eyes and the bulging sad clown(ish) eyes have taken their last breath together. They met in class and discovered passion during off-campus tutorials. D pinned H against a Volkswagen for their first kiss. Sirens whizzed by, his backpack was still on, and it knocked the wind out of her, memorably. They didn't rush into anything but saw a lot of each other. She went to his shows. He listened to her stories. They sang karaoke in far-flung locations, made crank calls, and became familiar.

She recalls that being inside his love was not dissimilar to sitting in the "middle seat" on the subway: the available space was deceptive. He said the relationship needed to be snuffed because it was soot on the wings of his career. But their untimely ending also seemed to be caused by garden-variety stress and cowardice. It stemmed from his unconscious darkness. . .or fear about their significant age difference.

D is remembered for how damn fun he was to play with and how patient he was with her neuroses, large and small. She adored him more when he didn't have to schmooze or be the life of the party. H is possibly remembered for having a smart mouth hinged to an off-kilter sex appeal. A wearer of costumes. A woman that was hard to drive or leave on time with. Her feelings toward him were filled with electricity and tenderness. He recalls sometimes that tired him out. But periodically, he misses being tired by her.

They're survived by the term "dogballs" carved by D in the sidewalk outside the Brooklyn apartment where they fell in love, Mr. Scrambles, and the late Sunday evenings they used to wrestle in together.

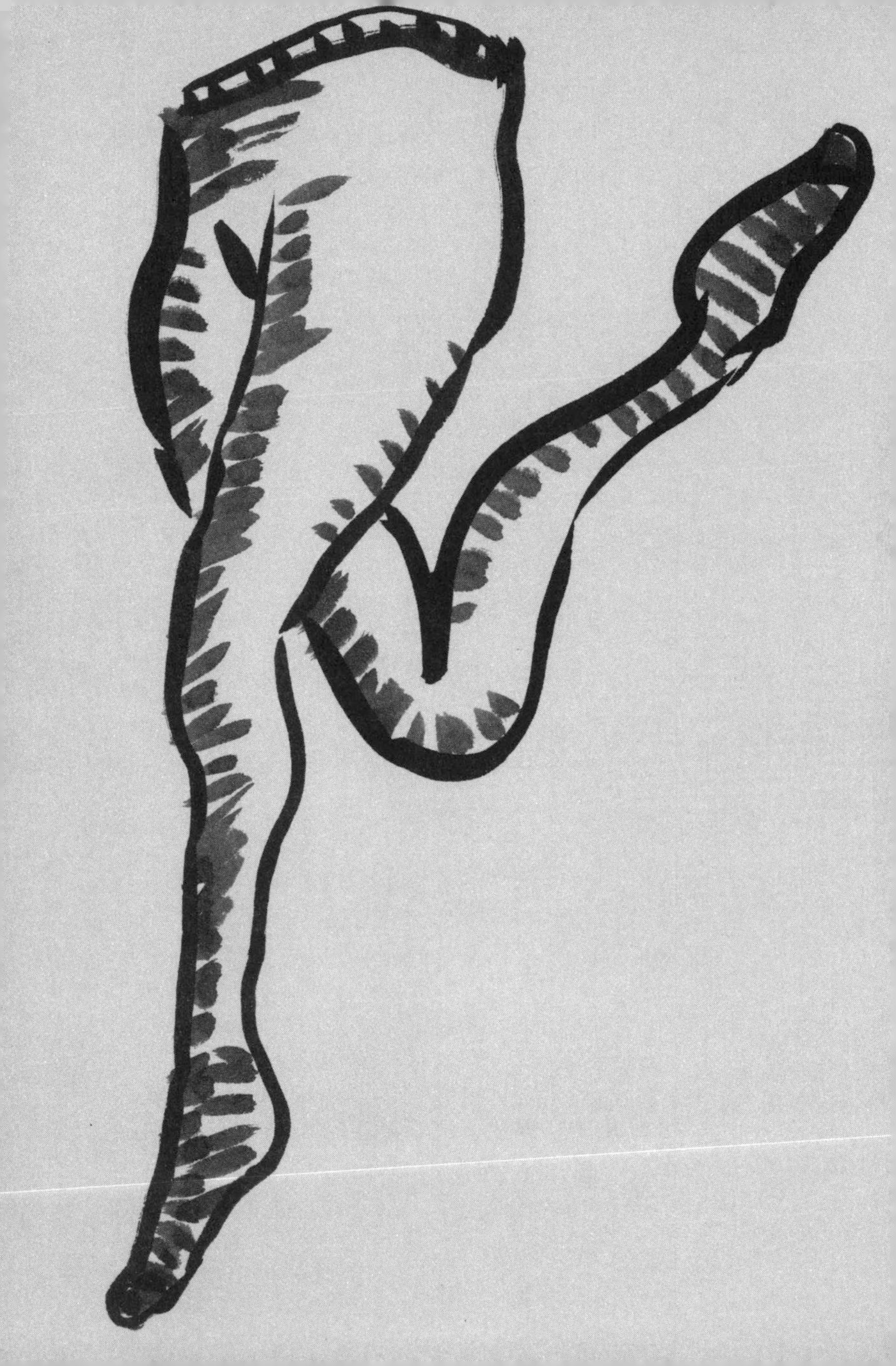

Cause of Death:
A 12-ounce can of beer

It's Simple, Really

Haley and Jeff the Asian
Born: May 5, 1998
Died: January 15, 2002

He told me Coors Light was his best friend, not me.

Cause of Death:
Time fears

Left Cross Lovers

Mary and Steve
Born: October 31, 1988
Died: April 5, 1992

Mary and Steve's four-year on-again, off-again relationship died suddenly of shock when Mary asked Steve if he had any intention of marrying her. He said he had considered it but was worried how he'd feel as time passed; Mary was eight years older than Steve, and he feared he might not be attracted to her as she aged.

Before their love succumbed, it had been steadily deteriorating despite utter denial. One night after seeing the movie *Thelma and Louise*, Mary drove over to Steve's apartment. When he opened the door, wearing a ratty blue bathrobe and a pukka shell necklace and looking like he'd been recently fucked sixteen ways to Sunday, Mary socked him in the jaw. After that, Steve often referred to her as Left Cross Mary.

Soon after the brawl, Steve went off to seek his fortune, working as a truck driver. He called Mary while in truck stops across the nation and professed his commitment to her from every state in the union except the one they both lived in. The farther Steve got away from Mary, the more desperately he wanted her.

Mary and Steve first met on Halloween night, quaffing a Guinness and gazing into the mirror behind the bar, into each other's green and bloodshot Irish eyes. There was instant recognition, and their fate was sealed. They danced every dance. He said, "You're special." She said, "I'll have another."

The now long-lost lovers will be remembered for spending an entire day whispering to each other, transferring cherry-flavored mineral water mouth to mouth. For singing a duet of Hickory Wind. For riding up Highway 1 from Monterey to San Francisco on a motorcycle, saying nothing, feeling everything.

Mary will miss the laughter lines around Steve's eyes and his wide freckled hands. Steve will miss Mary's glorious clavicle and all the rest of the best thing that ever happened to him.

There are no survivors
in this relationship, except
the memory, wry and sweet.

Cause of Death: Cowardice

Brett and Camille
Born: March 6, 2007
Died: July 31, 2008

On July 31st, 2008, the corpse of Brett and Camille's relationship was found in a ditch outside of Brett's Louisana apartment. The corpse— little more than a muddied lump of black and white matter—was slumped over on its side and slicked with tears, and was barely recognizable to those who knew the year-and-a-half-long relationship as a solid mold of two beings. However, friends of both Brett and Camille have come forward and confirmed that the body, while extremely degraded from what they had seen just a few weeks prior, was definitely that of their friends' relationship.

An autopsy of the remains revealed what had long been suspected but never confirmed by the couple: that Brett and Camille were each suffering from chronic illnesses that made the death of their relationship inevitable. Brett had been a lifelong sufferer of Cowardice, a condition that obliterates the spine and shrinks the testicles. It rendered him helpless against facing any of life's challenges, including standing up for himself, doing things that pleased himself and not others, or standing by his girlfriend when

"It was at that
moment that the last
traces of Brett's spine
crumbled (his testicles
had shriveled away
long before)."

she needed him. Brett inherited Cowardice, a recessive trait that is carried on the Y chromosome, from his father, a small-minded man who himself suffers from a condition called Hyperracism. Brett's father was able to conceal his Hyperracism for most of Brett's life, but his ailment spontaneously manifested the first time that Brett brought Camille, an African-American, home. He began having extreme bouts of anger and heart palpitations, and he insisted that the only way that he could lead a healthy life was if Brett immediately terminated the relationship, even going as far as to say that he would cut his son off financially if he continued to see Camille.

Brett felt that the only way to nurse his father's Hyperracism and his own Cowardice was to openly end the relationship and carry it out in secret. Somehow, he convinced Camille to go along with the idea. It seemed like the only reasonable plan of action at the time. Camille, by contrast, was considerably healthier than Brett, having never suffered from a day of Cowardice in her life. When her own father began showing signs of Hyperracism in the presence of her new boyfriend, she simply told him to "grow up and get over it" because it was her life, not his. However, it only took a few months of secrecy before Camille's health began to deteriorate. A pain in her chest occurred the first time Brett talked on the phone with his parents on one of their dates; Brett's father asked him what he was doing that evening, and he replied, "Nothing." The word pierced through Camille's heart and formed a tiny hole. For the first time in her life, she FELT like nothing, like something that shouldn't be seen or heard or acknowledged. The hole in Camille's heart continued to grow over time, eating away at the organ until it was nothing more than a thin membrane of hope that Brett would someday tell his parents the truth. She tried her best to fill it—with Brett's constant affirmations that he loved her, the support she eventually garnered from her father, or the fun she had spending time with Brett's really cool friends. When Camille moved to Madison for graduate school, she even tried to fill it with denial of the problem, and constantly told herself that there was really no need for Brett to tell his parents about her since she lived so far away.

The relationship was crippled from the beginning, and Brett's Cowardice ("But he's my Daaaaaad! And I can't get a job!!!") was a crutch that propped up Camille's sympathy for only so long. Their relationship suffered a fatal blow while Camille was home for summer break, when she and Brett attended a friend's engagement party. It was there that Camille bared witness to a truly happy couple, sharing the joy of their love with their friends and family. Camille realized that a relationship is much more than the bond between a man and woman, and when you date someone, you date the people in their lives as well. The next day, she urged Brett to come clean with his parents. She told him that if they were to grow as a couple it was important that the two most important people in his life know about their relationship. It was at that moment that the last traces of Brett's spine crumbled (his testicles had shriveled away long before). He told Camille that although he "really loved her," he still couldn't bring himself to cause a rift between him and his parents. Camille promptly left his apartment, throwing the remnants of their union in the ditch outside of his apartment complex before getting into her vehicle and driving away.

Survived by Brett and Camille's relationship is a stuffed animal from Camille's childhood that she gave to Brett to remind him of her while she was away. Camille was smart enough to take it back before she returned to school, and now it has a warm, soft place on Camille's bed—undoubtedly more comfortable than being repeatedly thrown into Brett's closet every time his parents would come to visit. Also survived is Camille's dignity and self-respect. They had been missing for while, but are now filling the massive hole that used to be in her heart.

In lieu of taking her to the bar and buying her "break-up beverages" or throwing her an oft-discussed "Fuck Brett" party, Camille asks her friends to honor the death of their relationship by living their lives as kind and tolerant people who get annual check-ups to reduce their risks of developing Cowardice.

"Camille promptly left his apartment, throwing the remnants of their union in the ditch outside . . ."

Cause of Death:
Wrong words

"You Want a Vocabulary I Don't Have"

Hava G and Eric R
Born: September 2, 1970
Died: September 4, 1975

I love forever, no matter what. Every person I've loved is still in my blood. Some of course take up more space. I could recite the whole list in my sleep. Eric R was one of the big loves even though he looked a little too much like Jimmy Carter. I tend to fall for anxious ethnics. The country matters less to me than the requisite nervousness, accompanied by being self-effacing, and impossibly funny. Not in that is-he-really-funny-or-am-I-imagining-it kind of way. More oh-my-God-he's-funny kind of way.

We were 20 when we met, a very long time ago. We were together in college but very different kinds of students. He used yellow markers and actually underlined relevant sentences. He also jogged. I fought against those two prejudices because he could talk in a mesmerizing way. 24/7.

"Eric R was one of the big loves even though he looked a little too much like Jimmy Carter."

feeling

I'd been waiting for a champion talker all my life. He was a true grand master.

His mother hated me for various reasons, beginning with her sense that I was altogether too tall. My mother loved him, even though he wasn't Jewish. He was cheap in general, but once he gave me a shoe box entirely full of great costume jewelry that he'd collected. But that's not why I loved him. It was mostly his words, and his elegant sentences. He said "plethora," a few times. He knew thousands of things I didn't know: trees, stars, names of birds. He studied linguistics, and Spanish literature. He was in graduate school when we broke up. He was in Boston and I was in New York and it seemed, oddly enough, like my words changed and his did too. He started talking almost exclusively about the Mayan hieroglyphs. I wanted to know how he felt. He told me how the Mayans felt. Mysterious and accomplished was what he said. I wanted him to tell me how he felt about me. More than the Mayans and the names of everything. He called me Hava, for no real reason. Hava, he said once, you want a vocabulary I don't have. And that was more or less the end. Although every year on his birthday, I write him a letter I don't mail.

IDEAS

Cause of Death:
He needed space

Young, Sexy Guy Dumps Aging Jealous Girlfriend

Foodie Funbags and Eli
Born: January 5, 1999
Died: September 1, 1999

For four months they traveled together to beaches, through cities, and under water through Southeast Asia. Life on the road was exciting, but upon returning home, her additional weight and climbing age and his newfound freedom drove the nails into the coffin.

"If they were to meet again and alcohol was involved, they would probably end up in the sack."

They met at a campsite in a foreign land while backpacking. Over several drinks and a raging campfire, their love blossomed, culminating in a romp in the steamy, mossy, nasty campground showers.

Missed will be the exciting adventures and desperate need for one another.

Soon forgotten will be the jealousy and feeling of wasted time experienced by both.

If they were to meet again and alcohol was involved, they would probably end up in the sack.

Disappointment only surfaces when one hears that the other is happily married.

There are no survivors other than the dignity she left with when she realized his request for space meant it was over, and she ended it.

Cause of Death:
A toxic dose of sanity

Finally, She Dumped Him

MD and NC
Born: March 22, 2002
Died: August 24, 2004

MD and NC—known for their not-so-subtle differences in age, religion, race, regional accents, and net worth—died after drinking a toxic dose of sanity. They were two and a half years old.

The relationship started with feelings of dreamy bliss—despite a horrible backdrop involving the suicide of a best friend. MD found NC's humor, touch, and eccentricity a great distraction, and they became close—quickly. The bliss eventually devolved into lies, boundary crossings, and a visit by the police.

They met at an art auction, hosted by a good friend of MD's date. A shared interest in artificial intelligence led to the introduction of NC and MD's date, and of course, MD.

MD will be remembered for the best Christmas gift of all time and for being a Doubting Thomas in NC's life.

NC will be remembered for his persistent optimism and genuine desire to do good in the world.

MD & NC are **survived by** a stock options trial.

"We met online, a dating site. Apparently we matched on the 29 dimensions of our personalities. If only there had been 30."

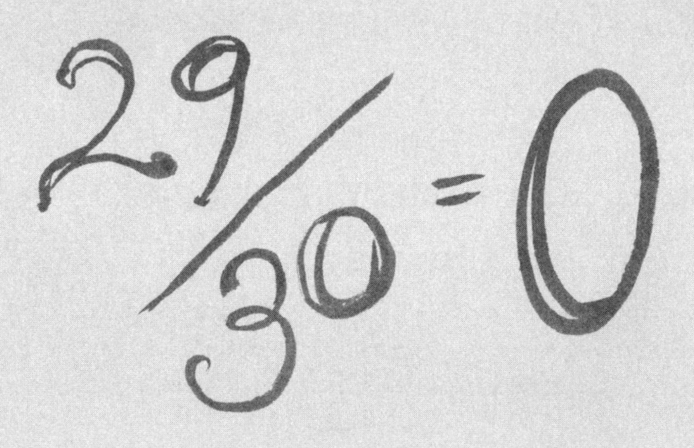

Cause of Death:
Narcissism,
stage 4

Expect More, Tolerate Less

S and N
Born: June 10, 2007
Died: January 5, 2008

Indicators the illness had metastasized: the continuing lies, the dismissal of any thoughts and feelings that were not his own; an inability to follow through on anything, big or small; a lack of self-awareness coupled with complete self-absorption.

We met online, a dating site. Apparently we matched on the 29 dimensions of our personalities. If only there had been 30.

The relationship began with eight months between our first and second dates. He had met his "soulmate" while I just wanted to be friends. However, after months of him persistently checking in with me, I agreed to hang out. We made plans to attend a summer festival.

He will be remembered for his generosity because although he is one of the most self-centered people I've ever known, it is not because he does not wish to be good to others. He will also be remembered for his lies, his temper, his manipulations, his empty promises, his lack of dependability, and his disinterest in my well-being.

I will miss the companionship, the laughs. He had a good sense of humor. Ultimately, I will miss the connection we could have had.

But I won't miss him.

I am disappointed that I didn't believe it when I first knew it.

He is **survived by** his mother, father, dog, two cats, two ferrets, a handful of "friends," a trail of lies, lots of enemies, bitter ex-girlfriends, and, finally, by his last girlfriend who is happy to be sad it's over.

Better Off Without You

"He will be remembered for being a complete and total

immature, insecure, fool."

Cause of Death:
Infidelity, lies,
and deception

Biggest Fool on Earth Screws Up Royally and Loses a Saint

AMG and GAG
Born: May 31, 1999
Died: August 1, 2007

Mr. and Mrs. Greg Allen G. died at the age of eight and a half years on August 1, 2007. They were happily married until Allison became sick with breast cancer and Greg couldn't handle it, becoming verbally abusive to his poor suffering wife. Greg began cheating on Allison soon after her long treatment when she was bald and recovering from the effects of the chemotherapy. Lying to friends and family, Greg kept this relationship secret, unaware that Allison was catching on to his rotten behavior. After finding sufficient evidence in January of 2007, Allison served Greg divorce papers and left him to the manipulative tramp he needed to "make him smile." This relationship was cremated by Allison, since she will not tolerate lying, philandering men. Allison is now at peace.

They met through Allison's best friend and Greg's brother who were married.

He will be remembered for being a complete and total immature, insecure, FOOL.

Allison will miss Sunday nookie, walking the dogs, cuddling on the couch, toasting their food, geek kisses, camping, and making English muffin breakfast sandwiches.

She will not miss being told that her job is a joke, that her butt looks different (this was during chemo, mind you), picking up that damn bathmat, cleaning up little shavings at the sink, or being told that she has no fucking idea about money.

She would be ecstatic if she never saw him ever again.

She is disappointed that he BROKE EVERY VOW and led her to believe he was ready for marriage, which means (HELLO!?) in sickness and in health, in good times and bad, etc., etc. It does not mean to stick your dick in some ugly whore you made out with ten years ago just because your wife was unable to have sex during CANCER TREATMENT. Poor you for having to jerk off for six whole freaking months. Glad you threw eight years down the drain, buddy. She hopes you are miserable forever.

Famous Last Words:

What goes around comes back around.

Cause of Death:
Complications brought on
by an In-N-Out burger

Burger Hell

J and V
Born: September 13, 2006
Died: April 20, 2007

J and V, a couple that died for a brief spell in March 2007 in Brooklyn, NY, expired for good in April 2007 in Reno, NV. J and V, who lived in Boston, MA and enjoyed floating down the streets of Cambridge, romantic cheese-steak dinners, and calling each other "wood mouse," survived for eight months. The ultimate death, due to complications brought on by an In-N-Out Burger in Nevada, was announced twelve hours later by J, who, after a long and broody flight HOME, during which V refused to play hangman or answer J's silly "would you rather" questions and told her she was "annoying" him, arrived at the dispiriting conclusion that they suffered from a terminal case of lack of goodwill.

The initial death, in late March 2007, succeeded a Sunday night viewing of an episode of "The Sopranos." J, a self-confessed nibbler, was feeling peckish and set off to explore V's kitchen. She found a box of blueberry-almond flavored yogi cereal and poured herself a bowl. V said nothing, but when the cereal was eaten and some milk remained and J spilled more dry cereal into the bowl, V flew into a fit of pique. "How much of that are you going to eat? That was supposed to be my breakfast for tomorrow morning!" Stunned, J put down the bowl and proceeded

"How much of that are you going to eat? That was supposed to be my breakfast for tomorrow morning!"

to cry. It was neither the temper that set her off nor the volume of V's words. Well, maybe it was, a bit. But most important, it was the fact that he didn't want to share. "If you really loved me, you'd want me to eat your whole kitchen," J said softly. V thought this over for a moment. "You're right. I'm sorry." They went to bed. They cried. Then they woke up and agreed to commit relationship suicide. They stayed apart from each other, exchanging the barest of e-mails. But then April rolled around, with its Tiffany-blue skies and attendant sweet smells, and the pair set off on a vacation. Well, a vacation of sorts: V's brother was getting married to a beautiful schoolteacher on a reservation in Reno, NV. V was feeling rather crabby, understandably—his younger brother had found eternal happiness, something that threatened to elude him until the end of time. To make matters worse, he lost $500 at the casino's blackjack table. But there was hope: an In-N-Out Burger awaited a ten-minute drive away. When the burger was attained, V let J have a bite. She liked it. A lot. She took another bite. And another. V couldn't take it anymore. "Enough already!" He snatched it back. "Get your OWN burger!" The sun continued to shine. The wedding went on. There was dancing. There was joking. There were hors d'oeuvres. There was probably some sex. It was probably enjoyable sex. But no matter. The deal was sealed. For the record: J and V are both slender.

Cause of Death:
"Nexus, Kingdom of the Winds" online game became more important than the person in the room, causing alienation, suffocation, irritation, and terminal disease.

Life in the Nexus Kills Couple

Peanut Butter and Jelly
Born: April 2, 1999
Died: May 18, 2006

It was the best of times, it was the worst of times. I wasn't ready for it, but she moved in anyway. Then we were Peanut Butter and Jelly: a great team, the model of a great relationship, two people who enjoyed nothing better than being together; until we weren't, it wasn't, and we didn't. She started playing That Damned Game and over time it took over her entire life from wakeup to sleep. I grew lonely and tired of being ignored. One day I realized she was talking on the phone to someone from the game in that sweet charming tone she used to use with me. I asked her

"Mostly, I do not miss being treated like a servant in my own home."

to leave, and six months later she finally did. Later, they moved in together.

Peanut Butter could be funny and sweet and charming. Peanut Butter could be cruel and abusive and narcissistic. Peanut Butter was addicted to That Damned Game, to drama. Peanut Butter was never happy unless she was unhappy. Peanut Butter taught me that I'd rather be alone than live in misery.

I will crave the backrubs. I will miss the road trips and the camping. I will miss those early days when I came home to someone who was happy to see me. I will miss having someone in my life I could talk to about almost anything.

There is so much to NOT miss: the noise, the drama, the phone stuck in her ear for hours, her crap everywhere, the farting, the selfishness and self-righteousness. I don't miss the feeling of being trapped, stuck, and responsible for everything. I don't miss the fear of being screamed at or the nightly rant. Mostly, I do not miss being treated like a servant in my own home.

Peanut Butter is gone. Peanut Butter had become someone else for someone else and I lost sight of her a long time ago. I hope that there will come a time when I can forgive even in the absence of apology.

The are survived by two cats each.
Also survived by a few friends who were
caught in the middle but didn't take sides;
and the many who did.

Cause of Death:
A nice girl looking for a
nice guy for companionship,
possible marriage

E-Harms-Me

C and V
Born: June 16, 2006
Died: October 16, 2007

We met on eHarmony because I know my man-picker is broken. So I decided to let the computer find me a man. We seemed to have a lot in common at first. He had a very nice manly voice on the phone. Knew just what to say. Was very good at keeping his real self hidden. Seemed very civilized and educated.

My first thought upon meeting him in person: "This guy is a dork." But I decided to give it a go. But as the months went by, I came to realize this man had a one-track mind: sex, sex, sex. He also suffered from hypochondria—his back pain was never-ending. He whined about it constantly. But the back pain never stopped him from wanting sex at all times.

I became resentful of him and his whiny ways. He did not turn me on and was totally clueless about foreplay, thinking a bottle of Astroglide was all he needed. One night when I declined to have sex with him, he threw a hissy-fit, so I made up my mind it was over. I left the next day, never to speak to him again. Since he did not call, I figured he was just as sick of me as I was of him. There are no happy memories.

Survived by his bottle of Viagra.

Famous Last Words:
You need therapy.

"But as the months went by, I came to realize this man had a one-track mind: sex, sex, sex."

50
Viagra
TM
50 mg

". . . and critiqued her behavior on all matters from drinking too much milk with her cereal to the finger she used when pushing the button on her camera (she gave him the middle finger)."

Cause of Death:
Mutual exhaustion

Death by Incompatibility

M and J
Born: October 30, 1999
Died: May 30, 2002

Brought together by the random forces of an online dating site, M selected J with a yellow highlighter pen upon printing out profiles of supposedly compatible females. The decision was made upon meeting in person that they were, indeed, compatible when M smiled and J felt she was struck by lightning. But lightning and mutual respect and adoration could not make up for the fact that these two people were, in fact, incompatible.

J found M too picky and difficult when he refused to slow down to her speed while walking, and critiqued her behavior on all matters from drinking too much milk with her cereal to the finger she used when pushing the button on her camera (she gave him the middle finger). M felt threatened by J's worldview and would not compromise on choices for co-habitation, for fear of becoming "yuppie" if he left his zone of comfort or cleaned his apartment. In the end, both felt emotionally alone despite enjoyment of similar activities, physical attraction, and the feeling that each person was pretty awesome in his and her own weird way. . .but they could never marry or reproduce without driving each other crazy.

"But lightning and mutual respect and adoration could not make up for the fact that these two people were, in fact, incompatible."

Cause of Death:
Lack of missing

The End

G and T
Born: February 21, 1987
Died: March 21, 2001

I don't miss driving him home from bars. I don't miss having bartenders calling me so I could drive him home because he had too much to drink. I don't miss waking him up when he would fall asleep in his car because he was drunk. I don't miss the fights. I don't miss his blackouts. I don't miss his friends driving him home. I don't miss the bars. I don't miss him falling down. I don't miss the concussions. I don't miss him trying to help everyone else and not himself or me. I don't miss the late nights. I don't miss the lies, the other women, and the ghosts from his past. I don't miss watching the same movie part over and over again or listening to the same song over and over again. I don't miss his need to kill himself. I don't miss his need to be the only reason I have for my life. I don't miss his need to be right. I don't miss being called a "good friend" when we lived together. I don't miss having sex being used as a weapon. I don't miss being afraid to answer the phone. I don't miss the worrying if he injured or murdered someone or was injured or killed

in a drunk-driving accident because there was no one to stop him from driving. I don't miss calling hospitals because he went missing till the a.m. because he fell asleep somewhere. I don't miss his memory loss. I don't miss cleaning, cooking, or washing clothes for him. I don't miss yard work.

I do feel bad about how he wasted his talented life. I feel bad that he got his wish for an early death. He had a great love for music, art, and poetry. He was extremely creative. Always sticking up for the underdog. He just lost his way and I couldn't help him find it.

Cause of Death:
Lies, mistrust, deceit,
and disappointment

Should've Seen the Signs. . .

Stubborn & Ambitious, Lazy & Narcissistic
Born: April 27, 2003
Died: July 30, 2008

They met on a sultry spring eve and danced together until dawn. She didn't expect him to call, but he did, and it changed her life—not for the better. In the beginning there was passion, trust, romance, endless nights of talking and making plans for the future. She should have known better when he proposed during a commercial break! The wedding was beautiful—small and intimate, surrounded by family. After the wedding, the problems became more and more apparent. She had left her job and changed cities to be with him. Then he got fired for punching a guy in the head—the rent was due, bills were stacking up, and life was suddenly not so perfect anymore. She found a great job while he continued to sit on his butt and do nothing for many months until she got fed up and made him get a job. She worked 65 hours a week to his 30 for almost 2 years, so he could stay home and play video games. She worked full-time and then some to pay off his repo'd truck and all his pre-marriage debt and still tried to love him despite it all. He started lying about everything. Over time, she realized that she wanted children, a home, vacations—things she would never achieve if she stayed with him, so she packed her things and filed for divorce.

He will be remembered for his blatant lies, impulsiveness, selfishness, complete lack of work ethic, and seriously flawed sense of logic.

At first, the loss of companionship was painfully hard to bear, but that feeling is long gone, replaced by confidence and perseverance.

If she never sees or hears from him again, she will be a much more pleasant and contented person.

No children were born of this relationship, and the couple's dog has made it clear that he wishes to remain in the care of his mother.

"She should have known when he proprosed during a commercial break."

"After the break up, he refused to talk to her and avoided her for a month, but did rehash their relationship with Buddhist nuns at the monastery."

Cause of Death:
He wasn't who she
thought he was

Glad That's Over

Ex-monk and Happily Single
Born: May 30, 2008
Died: July 30, 2008

They met at a monastery, which is a bad sign. They both lived there for awhile, celibate and very Buddhist. Out of nowhere, he became sullen and then gave a very moving speech about how he had to move out. He wasn't ready to commit to monastic life, and he didn't have the maturity or the self-awareness yet. He'd tried as hard as he could, and he just couldn't do it. He had to do what was right for him. She felt sorry for him, and when she decided not to be a nun and left the monastery, she spent some time with him. Soon, he was confessing his love, and she was bored and had been celibate a long time, so she went for it. They had a fun summer; some laughs, and meditated together. It was good for what it was. But soon, he started talking about how he wanted to travel the world with her and be

LOVE

her "spiritual partner." She believed it, and that was a fun idea too, while it lasted.

They planned to take a trip together, for her to meet his parents, and then to go to Buddhist teachings together. Two nights before they were supposed to leave, they had their first fight. The morning after the fight, an hour before she had to go to work, he broke up with her over the phone. He used the exact same speech he'd given when he left the monastery, which, of course, she'd also heard. Then he gathered all the things she'd left in his apartment and anything she'd ever given him, and left them outside her door while she was at work.

After the break up, he refused to talk to her and avoided her for a month, but did rehash their relationship with Buddhist nuns at the monastery. Then out of the blue, he sent her an e-mail. An apology e-mail. Among other things, it said, "Although I feel our parting was heartbreaking, I have tried hard to protect this remembrance from being tarnished by grief . . . I hope that you're doing okay, and secondly that I'd like to be friends again . . ." She politely replied that he had never been as important as he thought. Then she began using the phrase "remembrance tarnished by grief" and will always be grateful for that.

She thought he was a sweet kid before they dated, but during their relationship, she listened to him complain about everyone at the monastery and it was not peaceful. She sometimes misses the person she'd thought he was, before she got to know him.

After the break up, she took the trip anyway, skipped his parents, and hung out with her very cool sister instead. She had a wonderful trip. Since then, she's been much happier, and does not miss him.

If they ever met again, she'll thank him. He gave her a great break up story!

"She sometimes misses the person she'd thought he was, before she got to know him."

Cause of Death:
Acute nagging and disbelief

All Good Jokes Have a Punch Line

Poppy and LV
Born: October 11, 2000
Died: June 28, 2007

Like pork chops and M&Ms, Poppy and LV did not go together in the first place. They met in college, when LV fell in lust at first sight, and Poppy thought, "Eh, why not?" Soon, they were inseparable, sharing a love of everything on TV and a hatred for everything on the radio. After seven years, LV died of acute nagging, and Poppy passed on due to the disbelief that anyone in the world could actually be that stupid. Relatives and friends will miss their modern Abbott-and-Costello sense of humor, but are relieved that they no longer have to endure strained conversations and thinly veiled insults. Poppy and LV are survived by a love of "The Simpsons," "Anchorman," and college sports. In tribute, well-

"They met in college, when LV fell in lust at first sight . . ."

wishers will quote Bob Saget's monologue in "Half-Baked" and lines from "Chappelle's Show." Please send flowers and monetary donations to Bojangles and On the Border restaurants.

Famous Last Words:
How can we be lovers if we can't be friends?

Cause of Death: Alcoholism and selfishness

Len and Cammy
Born: August 11, 2001
Died: May 15, 2005

Although we were only legally married from 2001 to 2005, we had been together on and off since high school graduation. The relationship was rocky, to put it mildly, but I always convinced myself that despite the heavy drinking, staying out all night, emotional abuse, cheating, and lack of ambition or motivation to work, he truly in his heart loved me. One day I finally realized that since I was doing everything by myself, I may as well be by myself. The final straw came when my mother was hospitalized and he was too busy riding around drinking to even come by to see her. My mother had done more for him than his own parents ever did: helped pay for his school, bought him a car, etc. When I finally realized what a selfish, despicable person he was to others and not just to me, I was done. It was like he died to me.

I won't miss praying that he passes out before he comes to bed and forces me to have sex with him.

I'll live happily ever after. . .in spite of the fact that it took me so long to realize that he'll never change, regardless of what I say or do.

The relationship is **survived by** two great dogs . . . all mine because he's too irresponsible to even care for pets!

Famous Last Words:
Suck it, bitch.

"I will not miss his ability to suck the fun out of everything. He excels at it. He could suck the fun out of the word fun!"

Cause of Death: General meanness and complete insanity

Tom the Evil and Shelley the Stupendous
Born: June 4, 1994
Died: February 17, 2008

The reign of terror has ended!

It started badly and ended worse. Doomed from the start. . .

I didn't like him at all when we first met. He is the great chameleon, though. He morphed into someone I thought I loved. It was trickery at its best.

I will miss him being a great Dad for his kids.

I will not miss his ability to suck the fun out of everything. He excels at it. He could suck the fun out of the word fun!

I will now be forever happy. I am sad that he won't be around for the boys.

His boys will miss him dearly. That breaks my heart.

Famous Last Words:
Never regret something that
once made you smile.

S and N

Haiku Overdose

Born: June 22, 2007
Died: December 12, 2007

We liked each other
Then we didn't anymore
Now we plan three-ways.

"Two months after light goes out on the relationship, just as Americans have elected the country's first black President, and Bernard Madoff has been caught in the biggest Ponzi scheme in living history, undertaker shows up to collect remains of the romance. Upon his arrival, Girl is found in a snowy swirl of holiday hell and sadness. She is hunched over her computer, trying to resuscitate the relationship via construction of the Perfect E-mail."

Apparent Cause of Death:
Suicide

Actual Cause of Death:
Inability to Escape the Dark Shadow of Man's 20-year Marriage

Y. and the Girl
Born: July 6, 2008
Died: November 8, 2008

It started the same way it would end: at 2 a.m. on a Saturday night in Brooklyn, in a hazy cloud of alcohol. Distracting her from P., the pasty, undernourished writer the Girl had recently slept with, Y. seemed to fall from the sky.

She had landed with her friends at The Bar that night, trying to drink her way out of the rejection she was feeling. Y. looked to her like a like a dentist, a friendly guy you'd confide in but not necessarily want to go home with.

The Girl was in a cute, summery black dress with a head full of springy curls. He fell for her immediately. He loved the way she marched up to him and his friend. "My friend is flirting with a 25 year old and I want to give her space," she announced. "Will you talk to me?" He later told her

"She said . . . that he needed to date lots of people. He agreed. They cried a lot. They had sex. Then they cried some more."

that he'd never met anyone quite like her. She was sexy and smart and said whatever she was thinking; it made him feel like he could say whatever he wanted too.

He was from Wisconsin where men are raised to act like gentlemen. So, knowing she'd had too much to drink, he made sure she got home safely that night and insisted on sleeping on the couch.

The next morning, unashamed but shy, they wandered with her badly-behaved black lab/mutt, to find some weak deli coffee. They sat on a stoop. He seemed kind and cozy, like a big bear in grown-up clothes. She leaned into him. They made out.

They said goodbye so he could finish moving the last of his boxes out of his and his ex-wife's brownstone and into his new one-bedroom rental.

The e-mails and texts took a life of their own from there. They couldn't get enough of each other. Night after boozy night was spent in restaurants and bars or in his Eames chair listening to Scott Walker's "30 Century Man," looking out the window, smoking too many cigarettes. A creative director at an ad agency, he showed her documentaries about fonts and tried to explain why the band New Order was cool. And he got her excited about photography again, always sharing his latest photos, which he carried with him at all times.

The Girl liked that Y. loved fish tacos, that he didn't know anyone she knew, that he cried whenever he was moved or sad (which seemed to be on a nearly daily basis), and that he never made her feel guilty about taking taxis.

But Y. was confused. One month in, he told her, "Things weren't supposed to happen like this." He told her this again at the restaurant on his corner. He liked her too much. But he was barely out of his marriage.

When The Girl suggested monogamy after two months, Y. said yes and got all happy. Teary. But then, a few days later admitted that he was freaking out, "I mean it's not that I *want* to sleep with other people but I don't even want to be having this level of conversation. I mean I'm not even legally separated."

She started to keep a list of things she didn't like about him. 1) He didn't ask her questions. 2) He wouldn't see *Mad Men* no matter how much she said she knew he'd love it, and he wouldn't go to *Batman* because he didn't like to see movies in crowded movie theaters. 3) He could spend a whole weekend worrying about sunglasses and forget to watch the presidential debates.

Y. was wracked with guilt and anger towards his ex-wife whose rent he was still paying. All movie theaters and restaurants in a three-mile radius of his old place were off limits to the new couple. Even places that he and his ex had never been to together but that he thought she *might* one day go to were part of the ever-growing No Fly Zones. He could often be spotted looking furtively over his shoulder. He met all of her friends. She met none of his. She felt like a big dark secret.

Ultimately, Y. knew it was too soon for much of anything and didn't want to feel like he was disappointing someone yet again. And the Girl? Well, she was having a hard time stomaching all the push and pull. Nor could she stop the hovering biological clock from ticking.

"Unfortunately for all involved, the death turned out to be premature."

So, one Sunday morning in September, a week before he was to turn 39 and two weeks before she'd turn 37, just back from canvassing for Obama and full of conviction, she showed up at his apartment with the knife. She said she knew that he was still suffocating under the Dark 20-Year Shadow, that she needed more, and that he needed to date lots of people. He agreed. They cried a lot. They had sex. Then they cried some more.

Unfortunately for all involved, the death turned out to be premature. A week later they bumped into each other at his neighborhood coffee shop staffed by cute lesbians with lots of tattoos. Despite the protests of all of her friends, they drifted back together.

But it was just a matter of time. A month later, the Girl was trying her hardest to pin down weekend plans with Y. after he'd been out of town for a week. But, he was just "too busy."

So, just a few days after Obama's victory, with the whole country singing a chorus about all the Change that was possible, things came to startling end when Y. came to fetch the Girl from a going-away party and discovered her making out with a reporter who meant nothing to her. He was, appropriately, horrified. "Humiliated. Broken," he was overheard saying.

And that was all.

Despite her heartfelt apologies, Y. was resolute. He said he would never be able to look at her the same way again.

She understood. But it broke her heart. Though her friends later told her she was revising history, she was convinced she'd finally found someone who she felt at home with again. She tried to persuade him to see her, but he said no in a million different ways, a million different times. So the Girl felt sad and tried her hardest to leave him alone.

She decided she'd become an athlete.

And then a Buddhist.

And then one evening just before New Years Eve, she bumped into him. Some would say it was payback. Others would say it was good that at least it happened in 2008 and not 2009. In a tiny Tapas place right across from his apartment, Y. was in his black T-shirt with his soft stomach. He was on a date with a pretty well-scrubbed girl in a bright turquoise shirt. He looked relaxed and happy. And it seemed like the two of them had been together forever.

We'll never know more details than that. Nor do we want to. Because the Girl would like to think that she was special, irreplaceable. That what they shared was magic. Never to be forgotten.

May it Rest in Peace.

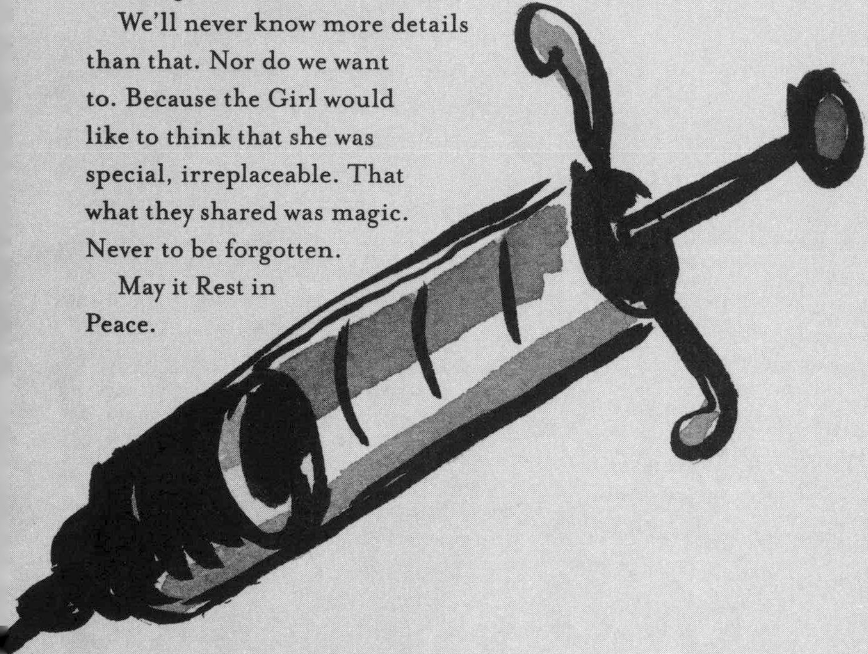

"Night after boozy night was spent in restaurants and bars or in his Eames chair listening to Scott Walker's '30 Century Man.'"

The Death of True Love

"How deeply did I feel for him? After several years I got married and named my son Tom."

Cause of Death:
To this day, I don't know
why we broke up

45 Years Later and I Still Remember . . .

Tom and Paula
Born: February 1, 1963
Died: February 14, 1963

We met . . . he had knee surgery . . . we dated . . . all of a sudden, he broke it off. His best bud couldn't figure it out. It absolutely broke my heart.

We met at a party. The girl hosting it got drunk. He didn't, I didn't. We talked. He told me he was having knee surgery. I visited him in the hospital.

He is the only guy I ever loved so deeply or felt so strongly about. (Of course, it could be because I was 16 and the feelings are just seared into my brain and I'm obsessive.)

I've missed that chance to have a relationship that started out slowly and built up to adulthood, marriage, and kids.

I will not and do not miss being 16 again.

I've not seen him since the day he came to my house, broke up with me, and left.

In those days it just wasn't right for the girl to "go after" the guy. And I apparently didn't have the guts to try to get him to be truthful with me. To try and see if we could fix whatever was wrong.

I don't know if he is still around or has married, kids, family.

How deeply did I feel for him? After several years I got married and named my son Tom.

Cause of Death:
Met too young

The Death of True Love

Christopher and Elizabeth
Born: March 7, 1995
Died: May 14, 1997

She found her soulmate too early in life. She didn't know how good she had it. She thought all men would treat her the way he did. She was dead wrong.

They were high school sweethearts who were soulmates. They loved each other very deeply, but were just too young, and life beckoned. If only they'd found each other ten years later.

They met in journalism class in high school.

The true, passionate love they felt for each other will live forever.

She will never forget the way he held her, or the strength of his hand in hers. She hopes that one day, in this life or beyond, they can be together again. She will never love again.

Cause of Death:
Suffocating from an emotionally
transmitted disease

The Love of a High School Romance Wasn't Enough to Last a Lifetime

The Seahorse and the Penguin
Born: March 12, 1996
Died: May 15, 2005

It started out simple enough: high school bad girl falls for honor student. We dated off and on for almost a decade, across states and oceans. He came to visit me in Hawaii; I moved from the sun and beach to join him in Illinois. We tried to take comfort in our familiarity. We were best friends, who loved to enable each other to despise life.

I will miss overindulging each other, on the couch with a pizza and a bad Liv Tyler movie. The ability to know

"I will miss overindulging each other, on the couch with a pizza and a bad Liv Tyler movie."

someone so completely. The unchallenged comfort we had all those years, no matter how much time spent apart. We felt like home for each other.

I can only hope that time will erase the memories of emotional indifference, the sarcastic retorts, the sick George and Martha-like relationship we came to foster. His overwhelming sense of arrogance, and my constant insecurity and need for reassurance, created a desperate love that only high school sweethearts determined to drain each other would endure for so long.

On a chance encounter, I would like to think the best of our abilities would shine and we could genuinely feel good for our successes and congratulate each other on our great lives without one another. What would most likely happen involves running to the nearest bathroom and revisiting one's lunch.

The inability to be friends should've subsided by now. We thought we could; we should've been able to. But seeing me drunk and tarted up falling all over strangers, and him keeping up appearances of a happy and successful life with his new wife, would prevent us from being able to complete a sentence to each other for years now.

The relationship is survived by many books. I hold a small collection of Hawaiian cookbooks gifted in moments of love and passion. He holds guides to the dsm4 I purchased for a birthday gift. He has a small collection of first-year anthropology books I pushed on him to "improve his cultural sensitivity." I have a small collection of his discarded psychology books that I never read in an attempt to understand him better.

Cause of Death:

He announced it was over

Disappearance

K and S
Born: June 21, 1989
Died: April 14, 1993

K and S were slowly drawn to each other . . . met over meetings, shared work, strong coffee, deep conversation. At a conference, by chance, they sat with John Cage, who pointed out that their names were all connected. That night, K and S began an intimate life together. They got mortgaged; planted a garden at a new "old" house; shared food with an ever-growing circle of friends. They planned for the future, lived in the present, and moved through the shadows of their pasts. They engaged in the living of a shared life that was pure in its stark difficulty and ongoing passion. They supported each other and were successful and prosperous. Then, four years later one early spring day, at a gas station not far from the dumpster, S announced it was over. He soon vanished. K wandered the streets, meeting the eyes of other mourners everywhere. She recognized them and became a member of the tribe that lives in groups of one. K slipped into the huge reservoir of loss that rests just beneath love and managed to keep breathing while moving deeper.

K and S are **survived by** a kickass garden and a deepened practice of loving.

Cause of Death: A comic misunderstanding

Pam and Jim
Born: February 2005
Died: October 26, 2008

A real-life Jim and Pam relationship passed away suddenly on October 26, 2008. Jim and Pam met at their workplace, a fluorescent-lit, windowless place of 9 to 5 drudgery. Pam was in a long-distance relationship with an aspiring English professor and Jim was trapped in a loveless relationship. They forged a bond over shared interests in books, music, and history and of course, a fanatic love of NBC's *The Office*. These shared interests helped pass the grueling hours. Jim and Pam shared many lunches together and long chats while they should have been working, shared laughs at the expense of their co-worker. Pam was oblivious to Jim's growing infatuation with her. He spent many hours at work and outside of it thinking about Pam and fervently wishing he had met her sooner, before he had become invested in his loveless relationship. Meanwhile, Pam was devoted to her aspiring English professor, who unfortunately had a wandering eye. After many difficult projects assigned by her demanding, yet psychotic boss, Pam left to take a job in the wilds of Western Pennsylvania. She wished Jim the best and hoped she could find somebody like him one day. Jim stayed at the job for a few months longer until he could no longer take the ranting of their demented manager. Besides, work was just not the same without his Pam. Once he secured a job at corporate, he left for an Eastern Pennsylvania City. But this real-life Jim and Pam's story arc was not yet over. After Jim had found his way out of his loveless relationship, he soon contacted Pam and so began their second act together. Pam by this time had finally ended it with the aspiring English professor (his wandering eye had wandered too far). They resumed almost daily conversation, often e-mail chatting with each other from their respective workplaces. Pam was glad to have her Jim back in her life and noticed his emails made the work day once again tolerable. Since this time around, Jim and Pam were both single they could admit their mutual attraction to each other. Plans were soon made for Jim to visit Pam. After a near-perfect weekend together Pam was smitten with Jim. Soon plans were made to see each other again, this time in the hometown of television's Jim and Pam, Scranton. After another great weekend, when Jim admitted he saw them married, Pam was hopeful that this long-distance relationship would work out. After a late night phone call from Jim revealed that he had loved her since the moment he met her and she was everything he ever wanted in a mate, Pam's heart swelled with love. However, she was confused by Jim's almost contradictory behavior towards her; dropping out of contact for days, never again an utterance of love. Pam chalked it up to Jim's stress at work and trauma from the past loveless relationship. A comic misunderstanding, one that belonged in the script of *The Office* is what ultimately undid this real-life Jim and Pam. It was revealed that Jim did not remember the late night phone call when he professed his love. He might have been drunk or even sleep-dialing. Furthermore he did not want to be in a relationship since the demise of his previous one, and he needed his freedom.

Survivors of this relationship include the NBC television show *The Office*, where television Jim and Pam are engaged to be married and have recently purchased a home in Scranton. Real-life Pam wishes them the best and wished the real-life couple could have been written in that story arc.

Cause of Death:
A clumsy stab to the heart
and slow loss of blood

Pabst Blue Ribbon

Carrie and Scott
Born: November 14, 2001
Died: March 17, 2002

There were good days, like when Carrie and Scott, both students at the University of Colorado, decided to drive to the Rocky Mountain National Forest to find, for no real reason at all, the bugling, blustering, battling elk. They didn't research the elk breeding rituals that well and ended up winding through the park, listening to music, and laughing. It was fall, and the sun was shining softly.

And, of course, there were bad days, like the time they were supposed to go out on New Year's Eve after Carrie got off her restaurant shift. Scott never showed up or even called. So a 21-year-old Carrie spent New Year's Eve in a pub with her coworkers, tears still wet on her eyelashes, pecking a friend at midnight who felt sorry for her.

Throughout those erratic months, Carrie practically worshipped Scott, but couldn't tell him or show him, and

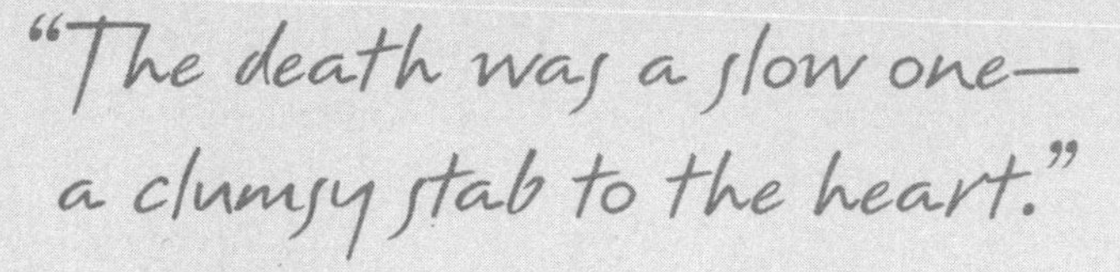
"The death was a slow one—
a clumsy stab to the heart."

in many ways she was right to distrust her feelings. In a sort of cyclical torment, she vacillated between two opposing truths. "It's my fault." "It's his fault."

Scott got sick of the push and pull, confused by Carrie's neediness and insecurity on the one hand and her rigidity on the other, and soon he started doing the pushing. Then it was over. And then it wasn't.

Their first, and only, "I love you" happened in a bar, after they broke up, after both of them lost close friends. It was the end of the night, and they stood near the empty bar, the bartender their only witness. Scott said it out of the blue. But it was too late. Or was it too little? "It must have been love," Carrie later said, "but I just didn't believe it at the time."

Carrie started drinking and crying a lot. Scott met someone else, a younger girl, short, sexy, tough—the opposite of Carrie. Then Carrie moved away, and when she came back, a kind of calm laid over their relationship, the kind that comes reluctantly and with time.

The death was a slow one—a clumsy stab to the heart, incredibly painful at first, and then a slow loss of blood, and hope, until the peace of the end set in.

Cause of Death:
Standard transmission

2,000 miles; September 11th; Lust

KH and SB
Born: May 15, 2000
Died: December 30, 2001

It really was a May–December romance, born in the office of a kids' theatre in Manhattan. She, adult child of hippies who'd left home (read: kicked out) at 15 and punctuated nearly every sentence with the word "fuck," was 30 when they made out at the sandwich bar in Times Square that rainy late spring (later: sex in the prop closet). He was 26, still living at home with his sweet-but-perpetually-stuck-in-1956 Catholic New Jerseyan folks, until he engaged in sex for the second time in his life, with KH, and subsequently moved to the West Village. This did not help assuage the circulating rumor that he was gay.

Everything was her idea: the sex (which preceded the romance); the romance; the prop closet; the move. She had been broken hearted, dumped by a former Mennonite who absconded to Mumbai to open a graphics sweatshop. And she had gone just a little bit nuts. He was goofy, blond, wide-eyed, inexperienced, and terrified. Humped like a jackrabbit, but she knew he had potential. Humping turned to love, as it often does. And thus the humping improved. Visits to his parents were excruciating for her, all that calling them Mister and Missus and sleeping in separate rooms. Weird for him, he countered, to visit her folks, where they shared a bed and called everyone by first names. Fighting, sure, but it was interrupted by adventures and long belly laughs; canoeing through the New Jersey Pines, despite his parents' objections ("where the mafia dumps bodies" was how his ma described the area); a day devoted to drawing a picture called "55 cats" for her mother's birthday;

"This was the happiest moment of her life: the drugged gray cat appearing under the arm of her smiling boyfriend . . ."

dressing as a cuttle fish for the Mermaid Parade. Often, he smiled with his mouth open and his eyebrows lifting; it was a look of pure joy.

He loved her cat, despite the allergy. She stabilized. He grew up a little. He read her bad short stories at that same sandwich bar and corrected her grammar with red pen. He drew funny, adumbrated pictures of squirrels. He let her cut his hair, even though one long wisp always dangled on the right side.

It was hard to leave all the little details that added up to love, but she applied to graduate school anyway. They drove to Utah together, leaving the cat behind to come later. He had never driven a stick shift, and, out of some vestigial manly instinct, refused to learn before they left. They stammered and creaked across the country. They fought in Kansas, outside Nashville, in Sedona, but they played Scrabble in each of the national parks and in the parking lot at Prairie Dog Town. The day he flew home, leaving her in the exhaust-pipe heat of Utah, she sobbed in the airport; it was August, 2001, and you could still accompany your party to the gate. A Mormon family had gathered there to bid goodbye to a son leaving for his mission: two years, disappeared in London. They huddled up, football style, and let out one collective heave of sadness. It shut her up.

Three weeks later, he returned to that same airport, her cat in tow. They were the last ones off the airplane. The steward was obsessed with the cat, he said, and asked if it was a Russian blue. This was the happiest moment of her life: the drugged gray cat appearing under the arm of her smiling boyfriend—no matter his suspicious fraternizing with the flight attendant.

She was due to return the day after her hometown erupted. You can't understand, he told her; he could see the smoke rise from his window, which used to have a view of the towers. Out there in Utah, it just looked like TV. Mormons in shirtsleeves and ties continued to knock on doors, looking for converts, and life went on unaffected by New York. Stranded in the desert, she took to sleeping with a drunken poet with thick fingers and an addiction to *All in the Family*. Just sex, she told him, but that was how it always started. Back in New York, he made out with her cousin's lesbian pal at a party, and slept with lover number three. From that he learned that he didn't care for one-night stands. By December, when he returned to Utah for a visit, they had cooled to each other. He smiled with his mouth closed now.

They parted amicably. She humped the poet. His one-night-stand turned to a relationship. The heartbreak came later, when she hit 32, boyfriendless in a suburb of Salt Lake City; the drunken poet had moved on, too. She realized too late the value of a man willing to fly a cat across the country, memorize all the two-letter Scrabble words, and grind the gears of a 1993 Honda he didn't know how to drive. Not gay, she told all her twisted-mouth friends who insisted that his insouciance made him so. Just a man not afraid to smile with his lips parted.

Cause of Death:
His love expired

Never Lasting Love

Martha and Chuck
Born: November 12, 2005
Died: September 4, 2008

A cool autumn's day had brought the two together. Chuck was a shy man with few words but had a kindness about him that Martha just couldn't resist. Chance meetings at a local restaurant had given them each other. Martha felt as though meeting Chuck was fate and after a few short weeks had promised Chuck that she would love him forever. Chuck was a hopeless romantic, an artistic writer of love letters, and would surprise Martha with flowers, just to see her smile. He too made a promise that he would be hers to hold until the day he died. Day after day the love they felt for each other grew stronger. With every glance and every touch Martha and Chuck grew closer together and were inseparable. On a warm summer's night two years later, while walking along the beach, Chuck asked for Martha's hand in marriage. With tears in her eyes, Martha jumped for joy, knowing that she would get to spend the rest of her life with the most wonderful man she had ever met. Days went on with complete bliss as Martha and Chuck started planning their lives together. The wedding was to be two years away in June, but wedding bells never rang. A year after Chuck and Martha's engagement Chuck fell out of love, a tragic event that was unforeseen. As Chuck told Martha that his feelings had changed, they held each other closer than they ever had and cried. Martha had given Chuck all of herself and vowed that her love for him would last forever. The night Chuck let Martha go, a part of Martha went with him. The vow that Martha made on that fateful night was one that will never be broken, nor ever tested. Fate had brought them together, a love that most will never have the joy of knowing. Only time holds the answers for the path that will lead Martha and Chuck to each other again.

Famous Last Words:

We will love each other again, even if it's in heaven.

he loves you NOT

Doomed from the Start

He Came Out of the Closet

Born: March 6, 1986

Died: September 11, 2002

'Nuff said.

Cause of Death: His still valid marriage certificate

Naïve and Bipolar
Born: October 2, 2008
Died: January 10, 2009

Bipolar is a cynic who enjoys horror movies, video games, and is in the military. Naïve is an optimistic, outgoing college student with a full-time job (and she doesn't really enjoy horror movies OR video games). Bipolar had just been left three months prior by his wife, Basketcase. Basketcase and Bipolar had been married *twice*, (which really should have been a red flag for Naïve). Basketcase left Bipolar on both occasions. This marriage had lasted only six months. Bipolar was devastated, and the type of person who liked to dwell on every little miserable detail of life. He tells Naïve he will file for divorce as soon as he is legally able.

Naïve, however, stays optimistic. Bipolar is a charmer, and at first, things are great. He is sweet, funny, and great in the sack, (which he got her into rather quickly.)

But soon things travel downhill. Bipolar becomes moody, detached. Finally, after leaving his house and beginning an eight-hour shift at work, Naïve is contacted by Bipolar over Yahoo Messenger. He doesn't want to touch her, or be touched. He doesn't want the relationship anymore.

A week later, Bipolar contacts her. He asks for "closure," saying that he would like to have just one more day with her. Naïve caves, and they arrange for Bipolar to come spend the night with her on her birthday, with the idea in mind that afterwards they would start over with a clean slate.

Naïve's coworker, Basically Perfect, asks her out. She has to turn him down because of the agreement with Bipolar.

Bipolar comes to visit Naïve for her birthday, he convinces her to be with him "one more time" and they "make love," which is something Bipolar had previously never acknowledged. Afterwards, Naïve is blissfully happy. Bipolar offers to go with Naïve on a road trip to visit her friends, and plans are made.

The day before the planned road trip, Bipolar flakes on Naïve. He says he is nervous about meeting her friends, and clarifies that they aren't a couple and he isn't her boyfriend. Naïve tells Bipolar she loves him (for the first

"The relationship between Naïve and Bipolar was short-lived, but jam-packed with drama, heartache, and lots of really good sex."

time) and can't deal with that not being enough for him. Naïve tells him she is going on the trip by herself and might talk to him when she gets home.

Naïve goes on a date with Basically Perfect. Bipolar is jealous and makes rude comments about her moving on too quickly. Naïve continues to see Basically Perfect, and tries to be "friends" with Bipolar. Basically Perfect is sweet, gentlemanly, and drives Naïve crazy.

Bipolar tells Naïve he loves her. He asks her to come spend the night with him. Naïve apologizes to Basically Perfect, telling him it isn't fair to him to be with her while she is in love with someone else. She goes to Bipolar. Things improve for the most part, and Naïve is happy again.

Naïve and Bipolar spend Christmas together, he asks her to move in with him for the third time. He is serious. She agrees. Naïve and Bipolar adopt a bunny together.

Bipolar tells Naïve he will never love anyone like he loved Basketcase.

The bunny passes away.

Naïve has noticed Bipolar becoming distant again. She tells Bipolar about her Coworker, Blonde Cunt, calling off her wedding. This launches a discussion about marriage, love, and relationships in general. Bipolar tells Naïve that because he is still legally married, she might not be able to move in with him after all. Bipolar has decided not to divorce his wife so that his step-daughter will still receive health benefits. Naïve is heartbroken, cries for hours, and decides to end things with Bipolar.

Naïve and her other ex, Still Awesome, drive to Bipolar's house to collect her belongings. Naïve says nothing to Bipolar, and he doesn't even look at her.

Bipolar is sweet, telling Naïve that he loves her every day. Naïve is hurting and confused.

Naïve goes to Bipolar's house to get the bunny cage. They sleep together and go shopping and out to dinner. Naïve is actually happy, and considers just being friends with benefits with Bipolar, since they got along so well. Naïve doesn't hear from Bipolar after she leaves his house.

Naïve gets in a small car wreck, texts Bipolar, and he responds asking about the condition of her car. Naïve sends Bipolar a few more messages, and finally he responds with "I just need some time." Naïve doesn't try again.

Bipolar sends Naïve a message, acting like nothing is wrong. Naïve asks what happened to the time he needed. Bipolar responds rudely that if she doesn't like it "oh well get over it." He said he needed space. Naïve gives him NASA.

The relationship between Naïve and Bipolar brought her nothing but pain. His mood swings, bad habits, and general dickishness made Naïve feel worthless, and his decision not to divorce his wife also rendered her effectively homeless. Bipolar made a huge deal of wanting to still be friends, then ignored Naïve after sleeping with her. This was what Naïve needed to finally see the truth. Naïve has decided to move on, and is ceasing all communication with Bipolar.

Naïve will most miss kissing Bipolar, his hugs, and the way he made her laugh.

Naïve will not miss Bipolar's extremely loud snoring, video games, and mood swings. But most of all she won't miss watching Bipolar slowly ruin himself.

Naïve hopes Bipolar will be happy mourning the sorry excuse for a marriage he had with Basketcase. Oh, and Blonde Cunt is pregnant. And Still Awesome is Even Awesomer.

"Then he converted to Catholicism and decided to become a priest, which was pretty much the deal breaker for her."

Cause of Death: He decided to become a priest

Dumped for God

Tempest and Father G
Born: March 15, 1994
Died: April 30, 2004

They fell in love, despite her being an agnostic and his being a born-again Christian (Southern Baptist). It wasn't a match made in heaven, since he thought that she was going to end up in hell one day. But she could live (or die) with that, because he was awfully cute and really knew how to throw a baseball. Then he converted to Catholicism and decided to become a priest, which was pretty much the deal breaker for her. Seeing him in a cassock two years later helped to make the break a clean and final one.

When asked for her final words about the relationship, she quipped: "I don't like to lose, but if I have to lose to someone, I am just glad that it was to God. I guess that I should just tip my cap to God and call him my daddy (or my big papi)."

Requiesce in pacem.

Cause of Death:
Lack of trust

The Relationship Expired As She Was Brushing Her Teeth

Jason and Meg
Born: April 1, 2007
Died: February 14, 2008

When she walked into his apartment she said she didn't want to overstay her welcome and would get home. He told her she never had to leave.

He had the most lovely brown eyes. She liked him for his hands.

She will always wonder if he went to Atlanta to meet up with that woman whose picture sat on top of the TV even after she asked him to move it. She will always wonder if he ever got around to telling Sarah he was dating someone new, since he held off telling Sarah for fear of

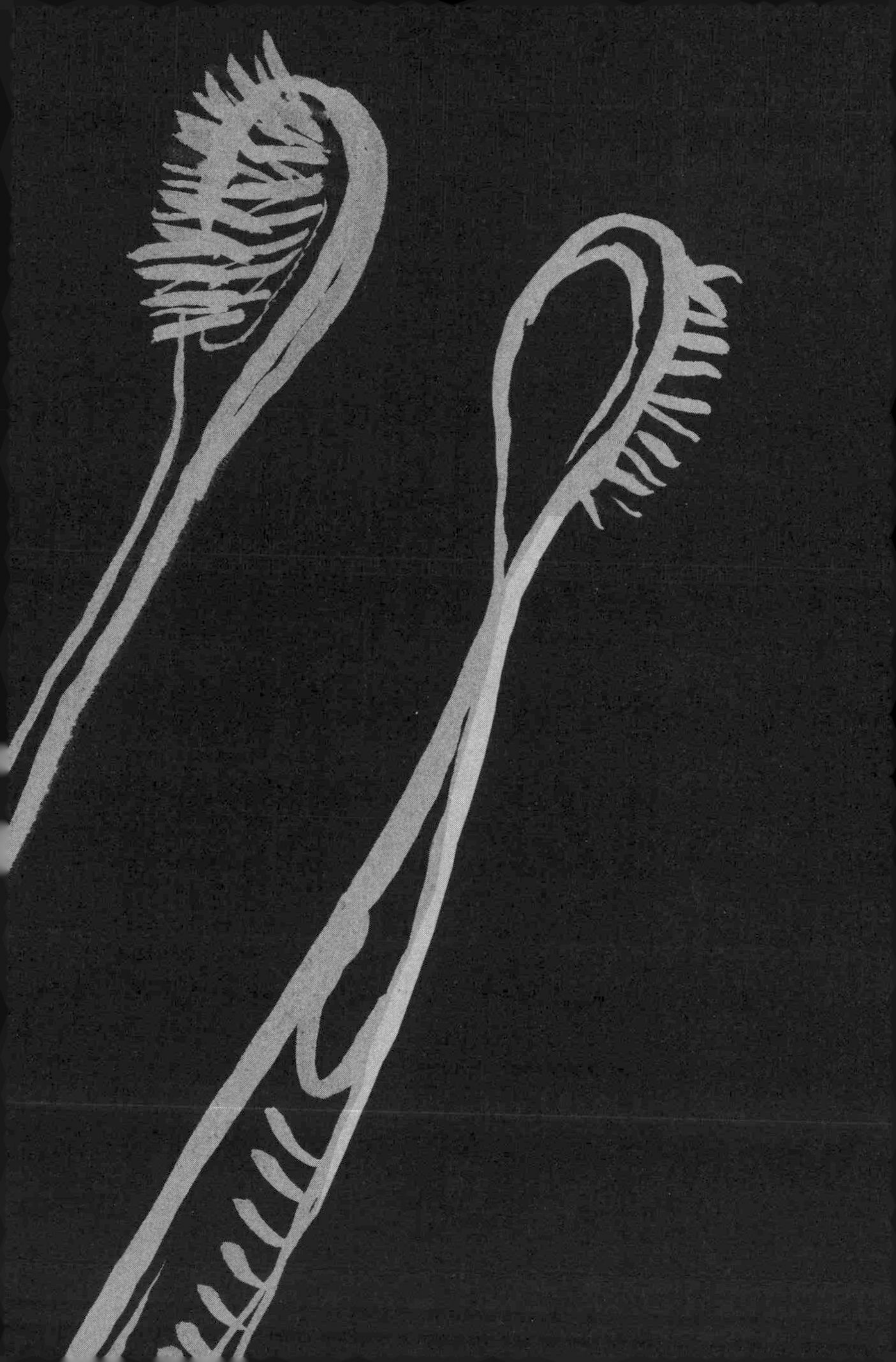

hurting her feelings. She will always wonder how he could take a lover as old as his mother.

He will be glad to be rid of her running hot and cold all the time. And he will be glad to get to bar hop around town again.

Donations of Miller Light (only the tall kind in cans) may be sent to his new Atlanta address.

The relationship expired as she was brushing her teeth. He stood in the doorway to tell her, "Actually, I did call her yesterday. Now don't get mad—I had to call her so she could get Ryan a job. Her best friend runs a bookstore, and he really needs a job, and I had to call her. I'm kind of, well, I'm kind of going to Atlanta to thank her. She's going to give me her old car, and I really need a car. Besides, she likes absinthe."

The couple is **survived** by two overstuffed couches, one dog, and a fresh bouquet of Valentine's Day flowers.

Cause of Death:
Fear, dishonesty

eDisharmony

Al and Grace
Born: June 18, 2006
Died: December 11, 2006

We were matched on eHarmony on June 18, 2006. We e-mailed and talked on the phone until we couldn't wait to meet each other. I felt like there was nothing we couldn't talk about. We met in an airport on Friday, September 23, 2006 at 5 p.m. We were both so nervous. I walked right by him and didn't even realize it. We couldn't stop looking at each other. While we waited in line to claim his lost luggage I finally grabbed him and kissed him. After that we couldn't stop. We continued to travel back and forth as much as we could. In November he was told that his kids were going to move to another state. I met his sister on Thanksgiving and flew up for his Christmas party.

I remember the last time I saw him. It was at the airport, Monday, December 11, 2006. It was the first time that he didn't call me after he saw me off. I remember him standing there waiting for me to walk out of sight. He looked so sad, but I didn't ask why. I still remember his last words to me: "Hurry so you don't miss your flight."

After that, e-mails came less. Phone calls began not to be returned. I tried to be the supportive girlfriend. I thought he was sad because of the kids. It wasn't until April that I learned he had a girlfriend on his MySpace page. Just a few days before, he had been texting me about his kids. I had no idea.

He led me on. Now a year later I still get an e-mail about once a month. With no explanation . . . only that he thinks of me.

"While we waited in line to claim his lost luggage I finally grabbed him and kissed him. After that we couldn't stop."

Cause of Death:

He got her, and then he was bored

Cowardly Jerk and the Girl Who Wasn't Good Enough

Born: July 17, 1998

Died: June 29, 1999

A classic, but tragic love affair. She didn't bring enough excitement to his life. She had too many needs . . . how dare she! The story is simple, really. The relationship is controlled by the one who loves the least.

He saw her working at a retail establishment. They were both 19.

He was a friend of a friend. She was involved with someone else. He convinced her to end the other relationship and give him a try. She finally did. She fell in love with him, while he fell in love with the chase.

He was her first true love. She fell HARD and fast. She had never been hurt before. She was naïve. A decade has passed, and the pain has lessened but is not completely gone. He left her when she needed him the most, during a difficult family situation.

She will always miss the wild adventures and sweet, tender moments of the relationship. She will miss planning their future together. She continues to miss his amazing body and beautiful face. Ten years later, she still misses his family.

She will not miss his immature humor or that he could never remember to flush the toilet. She doesn't miss his arrogance or that he knew he was extremely attractive. She will not miss him making her feel less than perfect.

Seeing him again would probably cause her heart to stop. She is happily married to someone else, but she still dreams of him at night.

There are **NO survivors** from this relationship. As he put it so eloquently on the last day of the relationship, "I love you . . . just not the way you need me to."

And with those words, it ended.

Cause of Death:
Complications relating to
terminal age difference

The Internet Age

Jane and Emily
Born: June 28, 1999
Died: September 30, 2004

They'd been together about five years—that is, if you count the first two, when they were "long-distance." Jane was a bored 16-year-old, who posted on an Ani DiFranco message board, asking for help making a fansite dedicated to the musician. Emily, a 24-year-old designer on the East Coast, offered to help. The two began talking over ICQ regularly, late at night. The Ani site was never made, and strangely, Jane never found it odd that someone eight years her senior would be interested in her and her Internet persona. The two had enough in common, she thought, and she didn't know any other lesbians, IRL. There was no harm in their e-relationship.

Jane's mom begged to differ. She accidentally intercepted e-mails between her daughter and this person she'd never heard of, who signed e-mails to her daughter from

"The two had enough in common, she thought, and she didn't know any other lesbians."

her graphic design company address with Xs and Os. As it was not yet so common to meet people online, Jane's mom called the company phone number listed in the e-mail signature to find out who this strange woman (clearly, not a teenager) was, and why she was sending virtual hugs and kisses to her daughter. She told Emily to never contact Jane again, so the prognosis was not good. But Jane was still bored, and still had her own computer and phone line, so the two continued to communicate. They met a few months later, when Emily flew to Illinois so the two could attend an Ani DiFranco concert (naturally).

They managed a couple more visits while Jane was still in high school, thanks to her friends from summer camp who lived near Emily and could serve as alibis. When Jane moved to Chicago for college, Emily requested her company transfer her there. Thus began a real-life relationship which consisted primarily of vacations, dinners out, concerts and movies, all of which Jane, a student, allowed Emily to pay for (along with most of their rent and bills). Jane, only 18, 19 and 20, naïvely thought that being taken care of was love.

But when she was just barely 21, and entering her senior year of college, Jane realized she had hardly lived a true college experience, and no longer felt like rushing home from class to her "old lady." Plus there was the issue of Sara, Jane's cute co-worker at the school paper whom she'd started spending all her time with. Emily went back east for a couple of weeks to give Jane the space she'd claimed she needed. During this time, Jane got to know Sara a lot better, in many ways. But she should have known from past experience to be more careful with her Internet passwords, as Emily grew suspicious, and found some incriminating MySpace messages between Jane and Sara the night before she was to come back, likely expecting a full recovery. She called Jane and told her to leave their shared apartment, which Jane, despite not having anywhere to go, was relieved to do.

After brief yet heated e-mails and phone conversations, Emily left Chicago, and two cats, for good in four days. Jane and Sara, who are the same age, have survived four years and vitals are good. The cats reside with foster parents.

"Jane, only 18 . . . naïvely thought that being taken care of was love."

Cause of Death:
Can you lose what you never had?

Sienna and Lewis
Born: May 23, 2006
Died: April 10, 2008

They met while Sienna was working in the hospital emergency room in Iraq. During the intensity of life in the desert and the reality that life is precious, Sienna finally gave in to Lewis's persistence. Lewis was gregarious, charming, confident, steady, brave, courageous, funny, and incredibly handsome. He was thoughtful and caring and kind, and when the hospital got too overwhelming,

he always knew how to make Sienna feel like the work was worthwhile and meaningful. He knew how to help her get over the images and the nightmares. Lewis was the most amazing person Sienna had ever met.

When it came time for Sienna to go home, Lewis came to visit when leave permitted. During one of the vacations to Mexico, Lewis proposed. Sienna had never been so happy.

Sienna received a Red Cross phone call on December 12, 2006. Lewis had been killed while on a mission. Sienna wanted to die too. In many ways, she did.

Fast forward 16 months. . .Sienna receives an e-mail forward with Lewis's name in the history. She panics and calls the sender of the e-mail to demand to know how that is possible. The sender eventually gives in and tells that the entire 'death' was faked so that Sienna would give up on Lewis. Lewis actually lives with his wife and three kids in Ohio.

Sienna will never really recover from this.

Survivors would be the new truck, the remodeled home, and the matching tattoo that Sienna now lives with because Lewis wanted them.

Cause of Death:
His lies

Duet for Clarinet and Violin: A Train Wreck in Two Movements

Clarinet and Violin
Born: September 12, 1961
Died: August 18, 2005

Movement I: After forty years we forgave each other.

Movement II: He lied to me and his wife. He persuaded his wife to go to counseling so she would realize that their relationship was dead. But (SURPRISE!) she thought they were trying to fix their marriage, and he didn't disabuse her because that would cause a scene. Meanwhile, he took me to Post Ranch and bought me pearls. He fancied himself a man of integrity.

Here's the best example of his integrity: For four years Clarinet reassured me that he wasn't really married because he was separated from his wife. Meanwhile, I spent

"He lied to me and his wife. He persuaded his wife to go to counseling so she would realize that their relationship was dead."

months copyediting his book, syllable by syllable. When the book was published, there was an entire paragraph thanking the cleaning staff at his motel and another paragraph thanking his "spouse" in particular for her help. My name ("and family") was buried in a list of helpful bit players.

Clarinet is a dishonorable coward who has always been motivated by fear of public censure. Forty years ago I was young, pregnant, and ashamed. Clarinet literally hid under my bed while my landlady reproached me for being a loose woman.

I will miss the illusion that love and forgiveness could redeem any personal tragedy.

Famous Last Words:

"You were right" is not an apology.

Cause of Death:
Inability to accept love

Loss of Innocence

Olivia and Jack
Born: January 5, 1988
Died: August 15, 2005

After almost two decades, Olivia and Jack finally laid their love to rest. It was a long, drawn-out death, both suffering from fitful delusions. Acknowledging the truth brought them to a timely end, allowing them to pass on in peace. They met during their first year of college when he gave her the book *School is Hell*, featuring a single-eared insecure rabbit he would soon come to resemble. For six months he courted her, climbing with her to treetops where they shared stories. He touched her hair tentatively, calling it graceful, before their first summer kiss in a moonlit meadow surrounded by granite mountains. They followed each other through the seasons of first love, watching the sun descend behind the Golden Gate Bridge, holding hands under a meteor shower on an island in the Sound. She realized, while he traced the lines of her collarbone with his fingertips, that it was the first time in her life that

"Making love only
made her miss
him more, and she
soaked many
of his shirts with
her tears."

she felt beautiful. She was his best friend, his first kiss and only love. But she felt threatened by his openness, believing herself incapable of relationship. She convinced him she would never fall in love with him, which made him cry into his soup. He retaliated by losing his virginity with her best friend. She couldn't acknowledge her jealousy or anger, even considered herself deserving of the betrayal for not loving him enough. But for each, the shame and insult of being denied left a deep and lasting distrust in its wake. Their letters and visits bridged them long distance. After graduation he took her flying, as at home in the sky as he was cradled between her hips. She felt intoxicated by his presence, but she couldn't ignore the evidence of his most recent affair. Making love only made her miss him more, and she soaked many of his shirts with her tears. When he visited the Other Woman and didn't return until morning, she left him with a note, never quite saying goodbye. The last time he appeared, unannounced after a long absence, he just stood staring at her empty cottage and felt the weight of all the time that had passed. After years of sporadic contact she wrote to him again, having never fully surrendered their relationship to the past. He hid her letter from his wife, who "wouldn't understand reminiscing about a pretty girl he loved at 19," and eagerly replied. Through a final year of clandestine letters, the excitement of their latent romance reignited until he disclosed, when pressed, that his marriage was the one thing in his life that really made him happy. The years of hope and heartbreak unraveled for her in that moment, and she admitted how misguided she'd been, projecting on

possibilities that no longer existed. She never heard from him again. He will be remembered for his brilliance, the way he made her laugh, and his revelation that how badly he had treated her was one of the biggest regrets of his life. She will be remembered for her desperate desire, her fear of being loved, and her courage to reconcile with the power of the past by placing it in its final resting place. The trials of their relationship will not be missed, though the delicious anguish of first love, forever imprinted on their souls, will always be cherished.

"But for each, the shame and insult of being denied left a deep and lasting distrust in its wake."

Cause of Death:
Obsession with himself

Venom is Poison for a Girl in Love

Snake and In Love
Born: January 28, 2005
Died: January 26, 2009

His name was Snake . . . his gorgeous green eyes enveloped her in poisonous venom and she fell into a daydream daze. What began with just talking quickly progressed and things heated up. They took their relationship to the next level when he told her he loved her. And she was swept away by his sweetness and the whiskey word . . . but couldn't see the trap she was falling into.

After a few months things started to get rocky and Snake started to change. His moods became violent and horribly strange, but she was blinded by her feelings for Snake. Ruthlessly, he broke it off and In Love was devastated. She ended up being rushed to hospital due to breathing problems (and severe heartbreak)! After a few days, they were speaking again. But Snake still didn't get how much

"Snake revealed himself to be a truly heartless bastard. Not only did he cheat on In Love—he put that whorebag on the phone to upset her even more."

she loved him . . . or how she paid for everything he wanted and visited his family on special occasions. And Snake never let In Love go places she liked: she wasn't allowed to go clubbing, out with her friends, or anywhere Snake didn't want to see her. She could also never understand why he never wanted to take her places—she was a very pretty girl. Towards the end of 2005, Snake revealed himself to be a truly heartless bastard. Not only did he cheat on In Love—he put that whorebag on the phone to upset her even more. In Love cried for weeks and turned to alcohol for support, but for some reason she still wanted the bloody Snake! After four months, Snake contacted In Love and wanted to start over. Stupidly, In Love fell for this and ran to his aid. She cooked the Snake breakfast daily and smothered him with tenderness. But it wasn't enough. Eventually she found out that Snake had been cheating again – this time with 25-year-old Life Ruiner! He pranced around town with Life Ruiner, taking her everywhere: (keep in mind, Snake had never taken In Love out for a dinner.)

In Love was lost, the clubs started to affect her and the alcohol that once blocked out the torturing pain wasn't enough. Six months went by and In Love finally found someone new. His name was So Sexy! He was great to her, appreciated her for who she was and spoiled her like crazy. It was a fantastic way to get over the lying Snake! But sensing she had moved on, Snake text-messaged In Love . . . all of a sudden so many memories penetrated In Love's mind and the feelings crept back. Stupidly, In Love dropped So Sexy . . . after that Snake wanted to see In Love every minute of everyday. She was so pleased with this, thinking someone had answered her prayers. But soon the storms rose once more and In Love was gushing tears again (as you may have predicted: Snake's moods had come back).

After so many cycles of the spurned lover feedback-loop, In Love realized it was time to move on. It took four torturous years, but she *finally* saw him for the venomous bastard he was.

She has since found someone who goes by the name of Perfect and is very happy.

Famous Last Words:

So over it!

Cause of Death:
Evaporation . . .

The Phantom Limb

Winona and Keanu
Born: April 7, 1994
Died: October 20, 2000

Winona met Keanu at church where they became friends. A palm reader and a Beck concert later, they progressed to young love, which lasted for years until the evaporation began. Keanu was a stable, loving, charming boyfriend who made all of Winona's friends drool, sending her romantic gifts and letters at college and spending every moment they could steal away together. But the long distance went on too long, and it was easier to be apart than together. It started on a summer break, when Winona was about to study abroad in Denmark. Before she left for Copenhagen, they took a walk on the beach and decided to take a break from the relationship while she was abroad. She looked down at their clasped hands and realized he was missing a finger. Where did it go? He didn't seem to notice it. Strangely, the relationship seemed to grow stronger the further apart they were. When she returned, they got back

"She regrets
that she
stayed in love
with a ghost
for so long."

together, but when he hugged her she realized that one whole arm was gone. Back at college, the relationship strengthened again through absence. Their intense love letters and late night phone calls generated a feverish heat of anticipation for the next visit. Yet, when Keanu visited Winona for her college graduation, his whole torso and both arms were missing. Who wants to have sex with a penis attached to two legs and a floating head? But the head was so gorgeous and it spoke such beautiful words, so she shrugged and kept mum. Again, he never mentioned the missing parts and they both pretended everything was perfect in their perfect relationship. When she finally returned home for good, she thought that things would get back to normal. However, the closer they were physically, the more he was slipping away from her. Every day it seemed like another toe, an ear, an ankle was missing. She clung to the parts that were still there and willed the others into being. Finally, Keanu told Winona he was leaving. "But I'm not a place you can leave," she said. "I've already left—haven't you noticed?" he said. When Winona looked up in tears, she saw that Keanu was a ghost—all of his corporeal parts had evaporated. So why did it still feel like he was there? For years after that initial breakup, Winona whispered to, cried for, and danced with the ghost of Keanu. She couldn't move on, because he was still lingering there. Gradually, she made the decision to live with the living instead of romancing the dead. But it wasn't that easy. He continued to haunt her, though appearing less and less often in her waking moments. One day Winona woke up and didn't think of Keanu at all, and then the next, and the next . . .

Winona won't miss dating someone thinner and prettier than her. She won't miss being with someone who hated anything that became too "mainstream," like breathing.

She regrets that she stayed in love with a ghost for so long.

Survivors include the Bert doll,
two heartbroken sets of parents, and
reams and reams of tearful letters.

Roll the emo music.

"Who wants to have sex with a penis attached to two legs and a floating head?"

Cause of Death:
Arrogance mixed with a
narrow-minded religion, assholitis,
and selfishness

Lost to Religion and Frats

DK and PO
Born: May 13, 2005
Died: October 4, 2008

He dated a girl for 3½ years and then broke it off with her because he said he could never marry a non-Jew. She offered to convert—but he wouldn't let her. This all happened right after he joined a coed frat. That was more important to him . . . you know, having friends that he would have to pay for and be their b!tch (only to never see them again after college ended).

He thought he was so awesome and everyone wanted to be his friend. In reality most people she knows hated him. This is a guy who thought it was okay to go out dancing with three girls and not invite his girlfriend because the girl that invited him only had one ticket left. There were good times

"She will probably kick him in the balls if she sees him again."

too, of course. But he never kept his promises—so she had more than a few issues trusting him. She believed he loved her, but his heart belonged more to himself.

She'll miss the times he was caring, like when he made her tea when she was sick, or when he sang to her. And he gave the best hugs.

She'll never miss (or will ever put up with again) his putting her down, making her feel worthless, and not inviting her to family events.

Thinking someone can change is a regret.

She will probably kick him in the balls if she sees him again.

Cause of Death:
Wanderlust in Thailand

Reading Your Male Mail Can Be a Bummer

Sawyer and Kim
Born: October 29, 1981
Died: February 26, 2006

His being a down-to-earth, real guy appealed to me. His ridiculous handlebar moustache aside, he was an attractive guy with a slightly crude, but dry and witty sense of humor. My only concern was our age difference; he was 13 years younger. He did drink more than I liked, but I was under the impression that this was from loneliness and boredom. I blindly assumed that it would improve once we married. Throughout the years we had our hard times like all couples, but my heart was in making it work and growing old together. My mistake.

We met when his younger brother was planning to marry my oldest daughter.

The worst part was the betrayal—his total lack of truthfulness with me. He never told me how he felt during our entire marriage; he lied, telling me I was perfect and wonderful. Right up until two days before he dumped me he sent me flowers and a card for Valentine's Day.

I'll always miss my friend. After so many years together, there was a very comfortable, easy way between us that I will always miss.

I won't miss, even a little, the bodily function noises and odors. . .the drinking and snoring. Having to live my life by his schedule.

I'll be just fine if I never lay eyes on him again and that most likely is what will happen since he's in Thailand.

The worst thing was his gutless method of telling me via e-mail that he wanted a divorce, when he had never even told me he was unhappy and when he'd been home on leave for a month just three weeks prior. As time proved, my first instinct that he was cheating was correct, so with all the lies, I was so disappointed in him as a person. I had always trusted him and believed him to be a very honest man. The disillusionment was overwhelming.

I most definitely will survive and never again believe the line, "I want you to be happy."

Cause of Death:
Kierkegaard

The Curse of the Fog of Angst

Celia and Hachiko
Born: May 26, 2005
Died: September 19, 2008

Hachiko and Celia died Thursday, September 19, 2008 in lower Manhattan after a protracted illness; the ultimate cause of death was Kierkegaard and Hachiko's mealy core. During their life together, the couple was best known for their shared workplace, messy and slightly drunken dinner parties, Hachiko's longing for the past and future, and Celia's waiting for him to land in the present.

They first met through work, but pointedly ignored each other in a series of early encounters. Then, the night of her 33rd birthday party he walked her home in the dark and told her about growing up in a provincial town in Scandinavia and the hicks who bullied him. A week later they kissed for the first time under a massive umbrella and

"He would wake up morning after morning crying about his ex-wife and Celia would try to comfort him."

later biked through the empty Brooklyn streets to his studio apartment, Celia perched nervously and excitedly on his handlebars. The next day Hachiko sent her a text message about listening to avant-garde music, poetry, and thinking of her. She wondered if it mattered that she knew nothing (and cared even less) about his tastes, but was charmed anyway and fell quickly in love with the vague way he moved through a day, his sideways cat-eyes, his non-linear beautiful mind, and how he was a secret goofball under all the Russian documentaries and rolled cigarettes.

Sometimes it felt like they lived in a magical fort. Hachiko loved how safe and held he felt in her presence, but he also liked getting a bit dirty with her, late-night in bars. Mostly they were kinder and gentler with each other than they'd ever been with any other partner.

Celia took classes to try to learn the hot-potato-in-the-mouth language of his country. They traveled there together and wandered in the forest, picking chanterelles, sweating in the sauna, and eating carefully prepared Scandinavian feasts. It broke her heart open, but fogs of nostalgia and longing began to descend on Hachiko. He would wake up morning after morning crying about his ex-wife and Celia would try to comfort him. It didn't help. He pined for his writer-school poet years of drunken late-nights of "dissecting and analyzing text." She found all that pretentious and annoying, and he would float further away, click-clicking on his computer in their Brooklyn kitchen dreaming up films to direct, poems to write, and projects to create, while she wondered when he would notice how sad she was.

They fell ill in May 2007 and began to disintegrate in a predictable way, when she said she wanted some sort of a commitment. A year later, long after he'd moved out and started dreaming about her instead of the ex-wife, he told her he wanted kids and commitment too. For a few weeks they gave it a try again, but quickly his mealy core undid him and he told her, "Maybe I'm like Kafka, Pessoa, and Kierkegaard. None of them could marry because of their work." He was poking fun of himself a little, but was also slyly serious. In the end it was a sad little joke on her.

Survivors include a tomato planter Hachiko built in their backyard out of wooden grocery pallets and their shared love of the overly earnest musical Hair.

"She wondered when he would notice how sad she was."

Cause of Death:
The relationship passed away in
a rainstorm of expectations, lies,
manipulations, and distortions
that reached its crescendo, like window-
rattling summer thunder, on
the walk home.

Choked

The Fictional Man and the Androgynous She-Devil
Born: September 27, 2008
Died: October 10, 2008

The She-Devil first acquainted herself with the Fictional Man by leaping onto his chest and dragging him onto the bed, not unlike a demon dragging one to Hell. Following the forward and incredibly modern advances made by the She-Devil, the Fictional Man found it virtually impossible to locate her. Speculation placed her in various layers of Hell. After weeks of evasion the She-Devil suddenly appeared to the Fictional Man again, wanting, well, who knows? His love? His companionship? His soul? (Most probably the latter). Still, the Fictional Man was only too happy to take the She-Devil to dinner and a movie, an offer she quickly accepted with measured jubilation. Dinner passed without much excitement. Midway through the film's contrived plot, the Fictional Man stretched out his elastic arm, and tried to wrap it around the She-Devil's shoulder. The She-Devil was politely revolted. (She may have vomited in her mouth).

The walk home was filled with brief, jagged dialogue—things better left unsaid . . . Then, the She-Devil offered multiple half-formed explanations for her words. The Fictional Man took it all in stride, allowing her to affirm the idea he'd developed that the inside of his mind was simply too cluttered for anyone but him to live in. At the end of the walk back, the She-Devil blurted out that she not only had a boyfriend, but that he'd soon be out to visit *and* that she needed a cigarette—so he'd just have to leave.

The She-Devil will be remembered for her ambiguity in a number of manners: her androgynous, short, and spiked black hair; her grimy attitude that hinted, barely, of back-alley fights and sexual inexperience; music taste too shallow to be eclectic, too deep not to be genuine.

The She-Devil will be missed for her diverse pseudo-personalities. Each day she pretended to be someone else. The variety was sexy.

The Fictional Man will not miss the She-Devil for her inability to differentiate between playful and bitchy activity. And the way the She-Devil gnawed at his mental stability.

From here on out, the She-Devil will most likely only exist in his 3 a.m. semi-consciousness fantasies.

True
LOSER
Always

I F***ed Up

"We were both young, both her age and my immaturity making us too young to have had any kind of relationship that wasn't built on anything but passion."

Cause of Death:
Overexposure and malnutrition

Beauty and the Beast

Jason and Nicole
Born: June 1, 1986
Died: August 20, 1991

Truly this was one of the darkest definitions of beauty and the beast: she was beautiful, articulate, intelligent, and prone to laughter and generosity. I was immature, and frustrated and angry at the world in general. I took these frustrations out on her through cruelty, manipulation, and psychological abuse.

We were both young, both her age and my immaturity making us too young to have had any kind of relationship that wasn't built on anything but passion. To be honest, as much as I kept screwing up in both our relationship and my military career (at that time), I look back and wonder why she stayed with me as long as she did.

We both needed each other, in that wild, passionate way that young people do. Her father had died in the year before I met her, and I was the strong alpha-male type she was missing since then. From my point of view, I had a beautiful girl as my own, the only good thing to happen to me in my life up to that point, and my paranoia of losing her made me possessive, yet unappreciative of her. I even cheated on her, which bewilders me even to this day as to why, as her beauty made me the envy of all my friends at the time!

The overexposure came in the form of my inability to control my temper, my will to dominate, and my desire to heap harm on anyone and to retaliate against the world in general. The malnutrition was the simple fact that besides commuting to see her every day after work, I contributed nothing of value to the relationship. She was always doing the "little things" to make the relationship work, yet I just took these things for granted, sometimes even teasing her about them.

Then came the day that makes me still cringe from embarrassment and self-loathing every time I think of it; the day I hit her. It wasn't even in the middle of some grand battle between us, or in reaction to any hurtful thing she said or did. It was the kind of blow one delivers to a dog that has pissed on the rug, the kind of blow that wasn't even physically harmful, but psychologically demeaning and hateful. The look on her face after that will forever be foremost in my mind as the symbol of the worst thing I have ever done in my life. When I remember Sara, the first thing I always remember is that look on her face in that moment.

That is truly a sad thing when one considers the years we had together, all the truly happy moments that came with the good times, and the times we supported each other through the rough patches that life can throw at young people trying to define their places in the world.

That thought even overrides thoughts of the last night we spent together before I went off to the first Gulf War. She did not expect me to live through it, as she knew I was in a front-line company and we had talked extensively about my will, life insurance, and other last business to conclude before I was deployed. She did everything she could to make sure it was the most memorable night of my life, despite everything I had done to her during the past four years.

She broke up with me shortly after my return. Eight months of separation from me had taught her what it was like being without me. She had gotten used to not being in an abusive relationship, had already prepared herself by then that I could have died at any time while gone. She had grown in that time and became a strong, independent woman, and I am proud of her for that.

Nicole, if you ever read this, I'd like you to know that although it may have taken me a while, I have grown as well. The decomposition of our relationship in the barren

"I even cheated on her, which bewilders me even to this day as to why . . ."

sand of my personality fertilized it and made it a rich soil. I am not perfect, yet what grows in the garden of my mind has truly turned me into a better person. In short, you alone take credit for making me into the man I am today, a man who appreciates the world around him, as well as the women I have met since then. My wife knows that when we have a daughter, naming her after you will not be a symbol of longing for the past, but an appreciation of how pivotal my relationship with you was in making me the man I am, and a reminder to never stop appreciating the woman I will share the rest of my life with. My only regret is that it cost you so much pain to teach me this.

Mit viel Reue, Sehnsucht, und Liebe (With much regret, longing, and love)

—J

Cause of Death:
Neglect, restlessness, fear

Lack of Intention

Orion and George
Born: April 30, 2006
Died: June 7, 2008

At first, they flirted without intention. They talked fancifully about teaching one another things neither one of them really wanted to learn: yoga for him, cooking for her. He would come to where she worked and pretend to shop. One time she was wearing roller skates and a yellow t-shirt. Tight. She closely resembled his favorite rock star. It drove him wild. He was spun. When he learned she might be moving away, he was quick to make his move. . .it seemed safe. He quickly grew accustomed to her smell and to the feel of her skin, and especially to the unique and beautiful way they made love. . .the perfect fit of their bodies, and the way he could toss her around by her tiny waist, which she loved. He gave her a lovely silver promise ring, inscribed with a quote by a well-known Sufi poet. She was moved to tears. He gave her all he could, and she loved him almost immediately for it. She lapped it up like a hungry kitten going at a bowl of milk for the first time. She did not realize he was already terrified. What would he do when he was out of romantic little tricks? Out of surprises? Out of money? My God, she would someday know he was merely a fraud!

So he began to despise himself for failing her before it had even begun, which of course accelerated the process. But she surprised him. She stayed, and the more he failed, the more she loved him. . .for a while. But she did want more, and sadly, once he found himself and gave himself completely and utterly, he found she had slipped away quietly, into the night, into the arms of another charming young man. She murdered it without real intention, and without warning, but it took a long time to die. In fact, it could be called "The Undead". . .a zombie relationship.

He was devastated. "I told you so!" said his cruel and unyielding ego.

She will be missed for her childlike smile, her aching desire to be held, and her impromptu mini recitals of ballet in the kitchen. He will be missed for his real and true concern, his black humor, and for the sex.

There are **no survivors.**

Cause of Death:
Hanukkah Harry

Santa Lost That Lovin' Feelin' in the Holy Land

N and Santa

Born: December 18, 2006

Died: March 6, 2008

N first spied Santa in a furry red sea of about 1,000 other St. Nicks as she was hotboxing a smoke outside a gay bar. He was participating in a drunken yuletide ritual called "Santacon." She was fresh from a violent de-coupling (and imminent divorce) from her childhood sweetheart. Santa floated over and bummed a smoke. Soon he was asking to put his mouth on more than her cigarette. She agreed to the meaningless debauchery because it felt like a reward for her first semester in graduate school. They necked all the way to her apartment. The next morning, they awoke to Santa-debris everywhere. N only gave him her business card before they said goodbye, thinking she wouldn't see Santa after Christmas morning. But he had already programmed his phone number into her cell phone when she was in the shower. That's how it went for 13 months—she was ready to write him off like the debris of a passing season and he never seemed to run out of presents.

She was hardened from her breakup and not ready to be serious. But he was happy to follow her around with his psychic mistletoe, pillowy lips, and adoring eyes. The following Christmas when N phoned him from her childhood home, complaining of bronchitis and overall malaise, he climbed in his Chevy Malibu sleigh and showed up on her doorstep to save her. Since he was younger and less experienced, it was difficult for her not to write off his generosity as naïveté.

In a few months, when he left for a trip to Israel, they were finally getting in sync. N had secretly decided to let herself love Santa. But it was too late. Santa experienced an epiphany in the Holy Land and traded his cap in for a yarmulke. "Hanukkah Harry" broke up with her over e-mail, saying they shouldn't see each other when he got back. They had nothing in common, after all. N misses the way he adored her and wishes she could have believed in Santa Claus sooner. Now she tries to avoid the winter holidays altogether.

Cause of Death:
Girls, girls, girls

Bright Lights, Big City

R and A
Born: March 1, 1982
Died: September 11, 1983

Our first real date together found us lip-locked at midnight, snugly seated in a lifeguard tower overlooking a ratty public pool in a small, sorry city—a real pockmark in the middle of Central California. A floodlight from a passing police cruiser chased us away. We ended up at a donut shop around 2 a.m. You: jelly, me: glazed. I remember you making macaroni and cheese in your underwear. And you looked impossibly good in cutoff jeans and a t-shirt. You snorted when you laughed too hard. You told inspired stories too. Like the one about your sister who worked as a nightshift nurse and had to attend to a patient in the emergency room who had willingly inserted a cup and saucer into his rectum. She later confirmed that it was part of a matching tea-set. Our friendship was effortless from the start—but while you envisioned a life together,

I thought of cities and places yet seen. At 18, I foolishly assumed there were other like-minded lovers for me out there too: girls, girls, girls. At 21, you knew better. Our pairing became a rare thing wasted. I moved. You stayed. I still take mine glazed.

"We ended up at a donut shop around 2 a.m. You: jelly; me: glazed."

love
RIP

The Modern Obit

A Q&A with Marilyn Johnson,

Author of *The Dead Beat: Lost Souls, Lucky Stiffs, and the Perverse Pleasures of Obituaries*

Can you give us a (brief) history of the modern obit?

There was a movement that started more or less in the 1980s in Philadelphia when the *Daily News* was trying to distinguish itself from *The Enquirer*. The *News* thought of itself as "the people's paper," so the editors found this wonderful investigative reporter (Jim Nicholson) and threw the challenge to him. He started doing these heavily reported, feature-length obits of ordinary people to sort of run against *The Enquirer*'s royalty and celebrity ones. It caught on—people really loved it. He got some attention and awards and other writers and journalists in various cities got into it. It spread to Canada, Australia, England. It was given a boost by *The New York Times* and other papers, after disasters would strike and the paper tried to bring those disasters home to people so it wasn't just like a collection of numbers of people to whom horrible things had happened, but actual stories about the people touched by the tragedy.

What do you suggest to someone writing a relationship-obituary? What advice do you have?

God is in the details, of course. You want something that avoids clichés. My guiding psychology when I write an obit is to try and find the turning point in a life—or in this case, a relationship. When was the point in the life when you could define a before and an after? What serendipitous or tragic circumstance happened that really changed everything? When I wrote an obit for Katherine Hepburn, I realized it was when she found her brother's body after his suicide. She managed to persuade her parents that he had not killed himself. That required a great deal of willpower, resourcefulness, and love. She was a teenager at the time. After that she found in herself the capacity to be an actress.

Some newspaper obituaries for celebrities or heads of state really investigate the people they're writing about and often use sharp language or include non-flattering details about that life. What do you recommend to someone who wants to document the failings of the ex without sounding vindictive or bitter?

You don't want to lose your audience. Another obit writer once told me: "I'm not worrying about being fair to the person or to the families. I'm writing an entertainment, not a history." Isn't that an interesting way to look at it? You're not just writing for yourself or a small audience. You're writing something that is going to go out there that you want people to read from beginning to end. It's not just a list of facts, it's a STORY. So you've got to have a beginning, a middle, and an end. The structure of an obit supports that because you are starting with the news of the death, then you shift back to the beginning of the life, you talk about what the person has contributed, and you talk about what he or she is leaving behind.

Do you think they should always be written in the third person?

No, I love writing that breaks form. I think that's fun; there's a lot of creativity. There's a whole movement now because of the democratization of the Internet where people are leaving behind obituaries that they themselves wrote. And they write them in the first person and they say, "If you're reading this, it means I've passed on." They are very powerful and some of them are quite funny.

Do obits have to be depressing?

I say that the death in an obit is a phrase or a sentence—the rest is about the life. If you talk to anyone who reads obits seriously, they're not gloomy people at all. It's like when you finish reading a story—you're rejuvenated. It's where the stuff of life is, what's left behind. Maybe if you were just reading "death notices"—those could be kind of hard to take. But the art of journalistic obituary in all of its forms is a celebration of the life that is left behind. It's the opposite of gloomy—it's often hilarious; you can't read some of them without bursting out laughing. People want the opportunity to laugh about death and I think most are ready to do this.

A Do-It-Yourself Relationship Obit Mad Lib

the beginning
the middle
the end
of love

Two young mortals met on a ____________ (adjective for weather) day.

Despite the fact that ____________ (your name) thought ____________ (your ex) was pleasant enough, it took ____________ (length of time) for them to finally hook up in the ____________ (location, noun).

____________ (You) took a liking to ____________'s (Ex) ____________ (physical attribute) but also appreciated his (or her) ____________ (nonphysical quality).

In spite of doing their best to play "hard to get," they took to calling each other ____________ (nickname) while no one else was around—and pretty soon, they ended up in bed.

Eventually, the two took a trip to ____________ (semi-exotic location). It was around that time the first "red flag" in the relationship emerged: ____________'s (ex's name) ____________ (choose the one that best applies: alcoholism, cheating, chronic unemployment, body odor).

In any case ____________ (you!) did his (or her) best to look the other way.

Instead he (or she) focused on ____________ (ex's best quality) and not the ____________ (nasty STD), etc.

But by __________ (date), the relationship took its last wheezing breath.

They attempted to resuscitate it by engaging in desperate make-up sex, but their union still croaked. An emotional autopsy revealed the cause of death to be __________ (reason you two split).

RIP __________ (you) and __________ (them).

The couple is survived by __________, __________, & __________ (noun[s]) but not __________ (deeply felt emotion).

Acknowledgments

I would like to thank the following people for helping to give birth to this book:

Kaari Pitkin, whose encouragement and support through this process has been invaluable—your friendship is an inspiration and a true gift in my life. Melissa Little for being one of the first to believe that I could bring the Web site to life, by sending the postcard that said simply: "Make it happen"; and for being crazy enough to invest in it even though you knew it was a shot in the dark. Web designer Stephen Strutt for taking on my nutty idea and creating exactly what I pictured (on layaway, no less)!

Editor Cindy DiTiberio for reading a blurb about the site in a magazine and envisioning a book . . . and for going through all the ups and down in the process and actually making it happen. Holly Bemiss who has been such an understanding and helpful agent—I'm really grateful I had you to turn to when I was knotted up and confused at certain key points. Author Marilyn Johnson for taking the time to discuss the art and elegance of the obituary. To everyone who visited the site and made the leap to writing their own relationshipobit. Justin Parks, aka "Xhard," deserves a special mention for allowing his work to be published here.

To my beloved colleagues at WNYC Radio —especially Karen Frillmann, Marianne McCune, and Rex Doane.

All my friends and family who've helped in countless ways—especially Esther Cohen, Anita Feder-Chernila, Susan Browne, Franco Stricklin, Amy Redford, Meg Dellenbaugh, Lauren Mechling, Stacy Abramson, Peter Mitchell, Bryan Smith, Ethan Millrod, Jennifer Marshall, Rick Fiscina, Rhea St. Julien, Molly Ferguson and Michele Villano.

And to Beverly Horan, a.k.a. mom, for your unending generosity and for embracing the fact that I've always been a little different . . . you're a friend to my heart.

STEFANO GIOVANNINI

About the Author

KATHLEEN HORAN is a reporter at WNYC News. She won a first-place award from the Associated Press for her feature story on the lives of food-delivery workers in 2005. Her work as also aired nationally and internationally on NPR, PRI, and the BBC.

Join the Online Memorial

Visit **www.relationshipobit.com** and put your deceased love affairs to rest by writing an obituary . . .